Henry C. "Hank" Smith
and the Cross B Ranch

Nancy and Ted Paup Ranching Heritage Series
Paul H. Carlson & M. Scott Sosebee, General Editors

Henry C. "Hank" Smith and the Cross B Ranch

The First Stock Operation on the South Plains

M. Scott Sosebee

Foreword by Paul Carlson

TEXAS A&M UNIVERSITY PRESS
COLLEGE STATION

Copyright © 2021 by M. Scott Sosebee
All rights reserved
First edition

This paper meets the requirements of ANSI/NISO Z39.48–1992
(Permanence of Paper).
Binding materials have been chosen for durability.
Manufactured in the United States of America

Library of Congress Cataloging-in-Publication Data

Names: Sosebee, M. Scott, 1961– author.
Title: Henry C. "Hank" Smith and the Cross B Ranch : the first stock
 operation on the south plains / M. Scott Sosebee.
Other titles: Nancy and Ted Paup '74 ranching heritage series.
Description: First editon. | College Station : Texas A&M University Press,
 [2021] | Series: Nancy and Ted Paup ranching heritage series | Includes
 bibliographical references and index.
Identifiers: LCCN 2020042308 | ISBN 9781623499679 (cloth) | ISBN
 9781623499686 (ebook)
Subjects: LCSH: Smith, Henry Clay, 1836-1912. | Ranchers—Texas—Crosby
 County—Biography. | Ranches—Texas—Crosby County—History. | Ranch
 life—West (U.S.)—History—19th century. | Pioneers—West
 (U.S.)—Biography. | LCGFT: Biographies.
Classification: LCC F392.C85 S57 2021 | DDC 976.4/05092 [B]—dc23
LC record available at https://lccn.loc.gov/2020042308

Cover:
Erwin E. Smith. *Ranching headquarters of Hank Smith, pioneer cowboy in the area. This house was built in 1877. Cross B Ranch, Texas*, detail, 1909. Gelatin dry plate negative, 5 x 7 in. Erwin E. Smith Collection of the Library of Congress on Deposit at the Amon Carter Museum of American Art, Fort Worth, Texas, LC.S6.460

Erwin E. Smith. *Hank Smith watering his horse, Cross B Ranch, Crosby County, Texas*, detail, 1909. Gelatin dry plate negative, 5 x 7 in. Erwin E. Smith Collection of the Library of Congress on Deposit at the Amon Carter Museum of American Art, Fort Worth, Texas, LC.S6.416

Back cover:
Erwin E. Smith. *Hank Smith, pioneer cowman of Texas, sitting on the brow of a hill in Blanco Canyon overlooking his ranch. Cross B Ranch, Texas*, detail, 1909. Gelatin dry plate negative, 5 x 7 in. Erwin E. Smith Collection of the Library of Congress on Deposit at the Amon Carter Museum of American Art, Fort Worth, Texas, LC.S6.490

*For Paul H. Carlson,
a mentor, influencer, and friend who never allowed me to give
up on this project and supported me until the end.*

Contents

Foreword, by Paul Carlson   ix

Preface   xi

Chapter 1: Introduction   1

Chapter 2: Becoming a Westering Man   6

Chapter 3: Seeking Opportunity in New Mexico and Arizona   27

Chapter 4: Hank Smith in Gray and Blue   44

Chapter 5: Hank Smith, Texas Entrepreneur   65

Chapter 6: The Cross B Ranch   87

Chapter 7: Crosby County's Most Prominent Citizen   112

Epilogue   137

Notes   141

Bibliography   153

Index   159

Illustrations follow page 64.

Foreword

Most often when we think of early cattle ranching in Texas, we think of big, open range operations, or we consider the huge King Ranch of South Texas, or we recall the giant XIT of West Texas. Ranches of immense size represent an iconic image of livestock operations in Texas and to some extent the larger American West. Visual memories of such ventures include a distinctive and youthful cowboy photographed on horseback watching a grazing herd from a distance or perhaps a drawing with similarly dressed herders on horses trailing animals north toward such Kansas railheads as Abilene or Wichita or Dodge City.

Not all ranching in the West or in Texas included such images. Indeed, most ranch undertakings were small, characterized by undersized landholdings, modest herds, limited financial resources, diversified livestock, few employees, and other features not found in expansive, nineteenth-century livestock enterprises. They often struggled to survive, and partly as a result they lack a good history.

As a critical component of its mission, the Nancy and Ted Paup Ranching Heritage Series includes works—such as M. Scott Sosebee's examination of the Cross B Ranch in Crosby County—about smaller, unfamiliar operations. Designed to provide fresh but distinguished books on ranching in America, the series stresses new research and modern analyses on the heritage of ranching, stock raising, and the men and women who toiled and struggled in establishing and maintaining such risky ventures. Whether describing a large or small, modern or long-standing enterprise, ranching history remains a multifaceted field of study and an agreeable topic on which to read.

Hank Smith and the Cross B Ranch is an impressive analysis of one of the smaller but significant ranches and its owner in West Texas. Smith's Cross B operation became surrounded by large-scale cattle ranches. Relations—both positive and negative—between the huge cattle empires bordering Smith's Blanco Canyon ranch provide insight into ranching struggles in the West, and they emphasize how Smith, in many ways a reluctant livestock man, expanded to survive.

Smith was an exception in many ways. A reluctant rancher, he secured initial ownership of his land and cattle in a peculiar but honest way. He preferred sheep to cows, but he always maintained cattle. He was among the first operators in West Texas to upgrade his range animals with purebreds—in his case using Hereford bulls to cross with his Longhorn heifers. He tried raising hogs, and at his home he planted a large variety of fruit trees. He spent more time on foot than horseback, tending his fruit trees and his garden. He used unique methods in managing and operating his ranch. In short, he was different, his landholdings were small, and he seemed more interested in gardening than in operating a ranch. Nonetheless, he succeeded quite well and died wealthy.

M. Scott Sosebee has produced a compelling and readable story of Hank Smith and his Cross B Ranch. Unusual in some respects, the book contains judicious insights and thoughtful notions and understandings. Sosebee explores big ideas and timely historical issues with broad parameters, and he places Smith and the ranch within those expansive concepts. His book, which uses documents and records unavailable to previous Smith biographers, represents an innovative study that, like other books in the series, offers an original interpretation in an agreeable prose style. It represents a valuable and modern contribution to the history of ranching in Texas and the American West.

—*Paul H. Carlson*

Preface

Countless people throughout history have made a mark, for better or worse, on the historical record, and scholars and chroniclers have documented the successes or failures of the many prominent figures. But there remain many people of local, regional, or even national importance whose lives are obscure to the generations that followed them. Such is the case with Henry C. "Hank" Smith. To many, Smith could seem to be the embodiment of the American dream, an immigrant youth who left his home in Germany to seek a better life in America and, starting with little means, combined skill and luck to become a successful entrepreneur. Through various business opportunities he acquired not only a degree of wealth but also recognition and respect from his peers as a merchant, freighter, stock raiser, and prominent regional and civic booster. Although he was a pioneer settler on the South Plains of Texas, his life and exploits were generally unknown outside of his contemporaries, close friends, and family.

Hank Smith's Cross B Ranch in Blanco Canyon was centered one mile northeast of Mount Blanco, ten miles north of present-day Crosbyton. But Hank Smith never intended to become a rancher. Born in Bavaria, he left Germany when he was only fourteen and traveled to Ohio to live with a sister. He then left Ohio after less than two years to seek better opportunities in the American West. In the course of his westering life, he worked as a teamster on the Santa Fe Trail, searched for gold in Arizona and New Mexico, served in both the Confederate and Union armies during the Civil War, and operated a freighting business in El Paso. In 1872 he left El Paso and eventually settled near Fort Griffin, Texas. While in Fort Griffin he married Elizabeth Boyle and opened the Occidental Hotel. Eventually, after foreclosing on a debt, he moved to Blanco Canyon and became a stock raiser.

The overarching theme of this biography will be contained within Smith's business activities and how his entrepreneurial nature became the driving force behind his movement west and the basis of his life. Smith's movement into the American West represented a model that was typical of many western migrants, a blueprint of entrepreneurship and settlement that was much more pervasive than the popular conception of a frontier pioneer experience. Smith saw little opportunity in Ohio and the Midwest and left in search of economic gain. In this regard, Smith's significance extends far beyond his life in Crosby County, an importance that makes him one of the prototypical, economically driven, "westering" men.

Henry C. "Hank" Smith
and the Cross B Ranch

1

Introduction

He was a tall man with a full beard and dark eyes. Although he loved to tell stories and make business deals, he was also prone to bouts of brooding silence that, to many, suggested an inner sadness. While he became an iconic figure on the plains of West Texas, a region known for its idolization of cowboys, Stetsons, and boots, he preferred to wear a broad-brimmed hat, overalls or suspenders, a simple white work shirt, and sturdy work slacks—with the ever-present pipe in his mouth. He was an able horseman, but one of his favorite activities was to walk his broad acres and, very often, camp on the ground in Blanco Canyon. One of the great joys in his life was tending to the fruit trees that lined the fence around his home. Most people have referred to him as a cattle rancher, but while he did raise cattle, for most of his life as a stockman he raised twice as many sheep as he did bovines. When he died he lived in a beautiful rock home; he left behind a considerable estate, but he had spent most of his life struggling to obtain a comfortable existence. He was often extraordinarily kind to his neighbors in times of distress, but he could also be stubborn and inflexible with some of these same neighbors when they were discussing buying or selling a horse or land. He and his wife Elizabeth raised five children whom they taught the value of education and hard work. He came to the United States from Germany and eventually settled in Blanco Canyon on the South Plains of Texas. He was Henry C. "Hank" Smith.

Countless people throughout history have made a mark, for better or worse, on the historical record, and scholars and chroniclers have documented the successes or failures of many of the prominent figures. But, there remain many people of local, regional, or even national importance whose lives are little known to the generations that follow. Such is the case with Henry C. "Hank" Smith. To many, Smith could seem to be the embodiment of the American dream, an immigrant youth who left his home in Germany to seek a better life in America and who, starting with little means, combined skill and luck to become a successful entrepreneur. Through various business opportunities, he acquired not only a degree of wealth but also recognition and respect from his peers as a merchant, freighter, stock raiser, and prominent regional and civic booster. Although he was a pioneer settler on Texas' South Plains, his life and exploits were generally unknown outside of the circle of his contemporaries, close friends, and family.

The scarcity of attention can be attributed, at least in part, to Smith's regional obscurity and lack of flamboyance. West Texas, and specifically the South Plains, was one of the last regions of the state to be occupied by white settlers, and myth and legend often mask its significance. Even today the region is plagued with the western myth of rambunctious and lawless cowboys, Indian raids, and cavalry campaigns, and identity of frontier existence. While these are, indeed, a part of the heritage of West Texas, they are a far cry from the whole story or even the most important aspects of the region's heritage. More important was the almost commonplace existence of families, such as Smith's people, whose everyday life was not "dramatic" in the classical sense, but one that adapted to an unfamiliar environment, developed a viable life and economic base, established communities, and became the true force behind the establishment of the region's present-day cities and economy.

Despite the mythic place the American West occupies in the American cultural mosaic, above all else, the West was a region of individuals, more often than not struggling to conquer and endure in a regularly harsh environment. It was such struggling people who transformed the American West into the dynamic region it is today, but they also helped to begin and perpetuate the imagined West of the modem era. Studies of such westerners can add to the existing work and help portray a more accurate picture of the West's past and present. Hank Smith was just such a westerner. This examination of his life will contribute to several themes in the current scholarly study of the American West. It will help answer questions on identity, economic activity, and the attitudes and motivations of the men and women who moved west in the nineteenth century. In many ways, Smith epitomized the "westering" man.

The overarching theme of this biography will be contained within Smith's business activities and how his entrepreneurial nature became the driving force behind his movement west and the basis of his life, particularly in his formation of the Cross B Ranch. Smith's movement west from Ohio represented a model that was typical of many western migrants, a blueprint of entrepreneurship and settlement that was much more pervasive than the popular conception of a frontier pioneer experience. Richard White, in his work *"It's Your Misfortune and None of My Own": A New History of the American West*, has speculated that opportunity in the nineteenth-century West was not as great as popularly believed, and that the West offered probably not any greater opportunity, as a whole, than the East of the same period. While that may presently be an accurate interpretation, it was not the contemporary perception. Hank Smith saw little opportunity in Ohio and left the Midwest in search of economic gain. In this regard, Smith's significance extends far beyond his life in Mount Blanco, Texas. He was the quintessence of the economically driven westering man. As Patricia Nelson Limerick has argued in her seminal work, *The Legacy of Conquest: The Unbroken Past of the American West*, the American West has a history grounded primarily in economic reality, in hardheaded questions of profit, loss, competition, and consolidation. Like Limerick's farmers, oilmen, and cattlemen, Smith's movement west and his activities were predominantly economically motivated, a search for profits and a better way of life for himself and his family. It was the search for profits and economic activity that was one of the most significant shaping factors in the American West, and Hank Smith was representative of such a movement.[1]

A study of Hank Smith's life will also contribute to potential answers in the broader questions of western identity. There are many possible answers to the question of what or who is a westerner. The West is home for many mythic figures and identities, but it is also nothing more than a region of the United States with no more uniqueness or significance than any other region. For many, the West represents lone heroic figures: the solitary mountain man, struggling homesteaders, daring cowboys, and the Native Americans. But for others, it is simply a locale, a place where ordinary men and women worked, raised families, and lived out their lives. The regional identity of a people is not monolithic, and it is expressed in a multiplicity of complex interactions between individuals, communities, and shared memory. The dichotomy between these two views accounts for the often very different views of the West and westerners.

Late in his life, Hank Smith began a memoir that is dominated by stories and tales of Indian battles, the romanticism of searching for a lost gold mine, and the adventurous life of traveling throughout the West. His stories, at first glance, seem influenced by the pulp fiction of the era, the larger-than-life struggles of a pioneer on the western frontier. This was Smith's life as he viewed it, and it was a life that reinforced the regional identity of the day. However, an examination of Smith's life reveals a much more complex picture. While his life did include adventures, it was also often mundane and more practical. He was also a husband, a father, a friend, and an entrepreneur. He was a major figure in the organization of Crosby County, served as both county commissioner and tax assessor, and at his death was a school district trustee. But Smith knew his audience and tailored his tales to meet their expectations of a pioneer existence. In many ways, Smith represents the dichotomy of the different views of the American West; this study will help reconcile and come to terms with these two divergent identities.

To his family, friends, and the present-day residents of the county he was instrumental in helping organize, Hank Smith is an extraordinary icon, and rightfully so. The fact that he was in many ways so ordinary, however, makes his life a valuable study. Some may take offense with this view of Hank Smith, a view that will eschew the romanticism and mythic characterizations of the West and make it more grounded in reality. In this vein, I offer a caveat: I come neither to bury Hank Smith nor necessarily to praise him unduly.

What this study will offer is a portrait of a man who is highly representative of what many westerners were and how their lives, or more accurately the misconceptions of their lives, helped to shape the contemporary perceptions of the region. In his migration to the West, his actions in the region, and his expectations for his life, Hank Smith was a very typical western migrant. Perhaps he did live an adventurous life of the Old West, but that did not make him an adventurous man. It simply made him a westering man.

2

Becoming a Westering Man

Henry C. "Hank" Smith's obituary in the *Crosbyton Review* on May 23, 1912, was full of praise and genuine sadness for the passing of Crosby County's most celebrated resident. It was no standard obituary buried in the back pages of the newspaper but demanded front-page above-the-fold coverage. It detailed how the German immigrant had come to the area for the first time in 1877, built the first ranch in Blanco Canyon, and was one of the key figures in the organization of the county. The lengthy article praised Smith as a man who "possessed just the right combination of attributes that usually make history-determination, rugged health and constitution, a natural supply of horse sense, and a desire to see the new world and to do things." It went on to note that he had accumulated "almost fabulous wealth . . . of cattle, broad acres, and above all, a picturesque old stone mansion snuggled away in the matchless Blanco Canyon." Even given the hyperbole of a turn-of-the-century newspaper article, it was apparent that Hank Smith was a beloved and iconic figure in Crosby County and the South Plains of Texas.[1]

Smith's life began as Heinrich Schmitt on August 15, 1836, in the small village of Rossbrunn in the northern portions of Bavaria. Rossbrunn lies in an area of gently rolling hills and rich river bottomland along the Mainz River, a prime agricultural region noted for its wines and vineyards. Bavaria had a rich and varied history beginning with its native Celtic tribes and their Roman conquerors. After the end of the Roman Empire, the Baiuarii (from which Bavaria gets its name) occupied the region; the Baiuarii were one of the five basic or stem duchies of medieval Germany. Irish and Scottish monks began the Christianization of the region, and it remains one of Germany's strongest Catholic districts.[2]

The agricultural wealth and the strategic position of Bavaria made it a coveted prize and a frequent battleground. Foreign armies often invaded, notably in the War of the Spanish Succession, the War of the Austrian Succession, the War of the Bavarian Succession, and the French revolutionary wars. In 1799, Elector Maximilian I Joseph united all the lands in Bavaria and aligned himself with Napoleon Bonaparte in the Confederation of the Rhine. With Napoleon's backing, in 1806 he became Maximilian I, King of Bavaria. But Maximilian abandoned Napoleon in 1813 and joined the allies against the French emperor; the settlement at the Congress of Vienna (1814–15) left him in possession of virtually all of present-day Bavaria. During the period of unrest that followed the Napoleonic Wars in Europe, Bavaria stood out for its relatively liberal government. The liberal constitution of 1818 lasted exactly a century. The revolution of 1848 did overthrow King Louis I, but the effects of the unrest in 1848 were relatively mild in Bavaria.[3]

Like much of Europe through the eighteenth century, Bavaria was a largely rural society structured along strict class lines. Agricultural production was the center of all village life, in work and recreation. It marked the seasons, the class lines, and provided the basis for village nicknames. In the early nineteenth century, the pattern began to change as the nobility began to lose their landholdings to the emerging agrarian capitalists. By the 1850s, approximately 40 percent of noble estates had bourgeois owners, and the new capitalist owners quickly became the village leaders. In essence, such leaders performed all the functions that the earlier elites had filled. They were the men who employed farm servants, hired the agricultural laborers, acted as the village foremen, took the lead when the local schoolteacher had to be kept in line, and served as sextons, church bookkeepers, and members of church councils. Under the new market system, land ownership, and the family reputations inextricably bound up with it, formed the basis of the village class system. But under this new system, the middle peasantry found it increasingly hard to maintain its landholdings, for the middle peasantry was more vulnerable to the cycle of debt, borrowing, and eventual foreclosure. Most such small, middle-class landowners found it only possible to keep going through intensified exploitation of family labor.[4]

The Schmitt family was one of these small landholders. The earliest records indicate that the Schmitts arrived in the area in 1680 when village documents recorded Johann Schmitt and his wife Anna living in the region. The couple were tenant farmers, and their descendants continued tenancy until Michael Schmitt, Hank Smith's grandfather, was able to purchase land in the early 1800s. Afterward, the Schmitt family continued as small landowners, but they constantly struggled to keep their meager holdings. During the early nineteenth century, supporting a family was difficult for small proprietors. In earlier periods, small farmers could depend on the village commons to graze stock or gather wood. But as commercialization of agriculture increased, land became nothing more than a commodity, and the new bourgeois landowners closed and fenced the commons, denying the subsistence farmer a vital link in his struggle to make ends meet. Thus, many small farmers had either to hire their labor out to larger operations or move into villages and learn a trade.[5]

Johann George Schmitt, Hank's father, was born in Rossbrunn in 1790 and, as a young boy, no doubt worked on the small family farm. His family called the young boy by his middle name, perhaps to differentiate him from several of his cousins who shared the name of their grandfather. But as he grew older, it became apparent to George that the family business held little or no promise for future economic success. As a young man, George had to depend on seasonal labor opportunities or other less regular jobs.

Eventually, seeing no other alternative, he joined the Bavarian Guard military force. He served as a corporal in the Bavarian Guard Regiment from 1814 to 1818 in Munich, and on returning to his home, again facing tenuous economic prospects, became a weaver, and moved into the village of Rossbrunn.[6]

George married Margaretta Herman in late 1818, and he and his wife raised eight children (four of their children died either in childbirth or infancy). George rose to a level of some prominence in the village, serving as a city commissioner from 1825 to 1840. Margaretta was a midwife in the village, and the family enjoyed some level of economic security. Heinrich (Henry) was George and Margaretta's eleventh child and their youngest boy. His parents gave him the name of an earlier child who had died in infancy and, by a number of indications, he was probably his mother's favorite. She pushed him to pursue an education and learn a trade. He was by all accounts a bright child who attended public school from the ages of six to thirteen.[7]

The Schmitt family lived in the family home in Rossbrunn, and young Heinrich enjoyed the carefree life of a German village youth. He helped his father with his weaving tasks and often ran errands for his mother. Rossbrunn was an ideal setting for a rambunctious boy. There were streams and a forest to explore, and Heinrich's numerous cousins provided playful company. School was no doubt an easy task for the youngest Schmitt male; in numerous accounts he was described as healthy and having a "good brain." Like most young German boys, he was expected to help with the household chores, but his favorable position with his mother kept him away from much hard work. She would often take him with her on her rounds as a midwife, and he became well known throughout the small village. The Schmitts' were devout Catholics, particularly Margaretta, who made sure that young Heinrich did not miss mass or confession, although she surely forgave her pet son any transgressions. Heinrich looked forward to a fruitful and happy life in the Bavarian countryside, probably with no thoughts of ever leaving Rossbrunn. But his idyllic life was about to take a serious turn for the worse.[8]

The Schmitt family's middle-class lifestyle was comfortable, but it was also fragile. Margaretta's work as a midwife allowed the family to send the children to school and have a few luxuries. But German society was undergoing tremendous change in the 1830s and 1840s. The working out of peasant emancipation left the countryside in a constant state of flux and turmoil, primarily because such arrangements often took years of maneuvering through the often-sclerotic legal system, and reform did not "give" peasants the land they worked. Instead, they had to pay or "redeem" their holdings. Like many former rural residents, George Schmitt had moved off his family's meager holdings and had chosen to enter the protoindustrial occupation of weaving at a time when the pace of industrialization was phasing out small textile firms in favor of larger operations.

While the family and village records are sketchy, it would seem that George Schmitt set up a one-person business in Rossbrunn and was beginning to face competition from larger concerns. Rural weavers faced a painful decline in the face of cheaper, more mechanized English factories, and although George did at least carve out enough income to keep his family clothed and fed, he had to be feeling the impact of competition. In his favor was the fact that he served an agrarian community that, at the time, still needed the services of a local weaver.[9]

While the economic survival of former rural peasants like the Schmitts was an often worrisome affair, simply staying alive was more frightening. Life expectancy in Bavaria during the 1840s remained desperately short; a quarter of all children died before reaching the age of one. Even if children survived into adulthood, they continued to face raging epidemics, such as influenza, cholera, and typhus. Perhaps the most tragic aspect a family faced was the loss of the head of the household; numerous German villages had a great number of widows trying to scratch out an existence for their children. In 1844, Margaretta and her children were faced with that very terrible prospect. On February 4, 1844, fifty-three-year-old Johann George Schmitt died. Although it is not clear exactly how he died, and Hank never wrote or spoke of his father's illness, Bavaria was stricken with a horrendous and devastating typhus epidemic in 1843–44. It is likely that this illness, or one of the many other epidemics rampant in Bavaria during these years, took the life of George Schmitt.[10]

Without her husband, Margaretta faced a grim future. Her income as a midwife was not enough to support her and the four children that remained at home. The oldest Schmitt son, Jacob (often referred to as Johann), also lived with the family but had a family of his own to support. He took over his father's weaving business but was in no position to support his mother and siblings adequately. The two oldest daughters, Anna and Margaret, were already married (Anna would soon leave for the United States) with families of their own to raise. The family had no real assets besides the family home or many prospects for the future.[11]

For the next seven years, the Schmitt family struggled to survive.

For young Heinrich, this period was no doubt one of great anxiety and apprehension. As his mother's favorite, he worried and tried to comfort the grieving widow. The young boy's formal education ended with his father's death, and he was forced to learn a trade to help the family. He was apprenticed to a machinist a few years after his father's death. The letter introducing him, from the priest who was the Relief Guardian of the Poor explained, "this boy is penniless, but is very bright and learns easily." Hank Smith did not choose to mention his apprenticeship in any of his writings, but later in his life in correspondence with Paris Cox, the leader of the Quaker settlement of Estacado near Smith's Mount Blanco home, Cox mentions Smith's "great skill with tools," suggesting that the young German lad did learn a great deal as a machinist.[12]

As the 1840s were ending, the family fortunes showed no sign of recovering. However, one bright prospect did seem to hold hope for some members of the Schmitt family, a notion that would have probably never entered their minds before the death of the family patriarch—leaving Bavaria for America. The Schmitts were not the only Germans of the 1840s and 1850s experiencing economic and social upheaval. In fact, the increasing number of emigrants was the most dramatic novelty of this period. Emigration from German lands was not unknown in earlier years, Germans having settled for centuries in eastern and southeastern Europe and the Eastern Seaboard of the United States. It was the volume that represented a transformation. The numbers began to rise in the 1840s, and the first great wave of emigration saw 1.3 million people leave the German states between 1845 and 1858. In the peak year of 1854, the total reached nearly a quarter of a million.[13]

Not all the rising numbers of German émigrés were facing the economic crisis the Schmitts were facing. Some had religious or political motives, but most did leave for material reasons. Like the now destitute Schmitts, the majority of the wave during this period was small peasants and craftsmen. While some Germans left for France, Russia, or overseas to Canada and Brazil, towering statistically over all other destinations was the United States. It accounted for 85 percent of all German overseas emigrants between the years 1840 and 1859.[14]

For the Schmitt children, the United States seemed an attractive proposition. Anna Schmitt and her husband had left Bavaria in 1847. Anna settled in Peru, Ohio, and wrote the family of the wonderful opportunity in America. In a letter home dated 1850, Anna extolled the virtues of her new homeland and urged her siblings to consider a move to the American Midwest. Many Germans viewed the United States as a progressive and dynamic nation, one with a growing economy and technological development. Rich fertile land could be bought for a low price, and transportation companies and land speculators made attractive offers. Individual American states also actively recruited within Germany to build their populations. German writers sent back glowing accounts of the United States and its land, governing ideas, and people. Gottfried Duden, in a widely distributed book, *Bericht uber eine Reise nach den westlichen Staaten Nordamerikas* (roughly translates as "A report following travel to the western states of North America"), described America as a land of beautiful scenery and unrestricted freedom. His depictions of the United States induced many Germans to come to America and left even more with fascinating visions.[15]

While it was an exciting idea for the Schmitt children, moving to America must have brought great pain to Margaretta Schmitt as she watched her family disperse. Immigrating to America was not an option for the now over fifty Schmitt widow. But by 1851, the reality was that she could no longer support her younger children, and the older ones were moving to America. Johann, the eldest son, made an application for emigration in late 1850, but it seems he never made the trip. His sisters, Margaret and Magdelene, had decided to accompany Johann to Ohio and continued their plans. Margaretta had a difficult decision to make. Heinrich was now fourteen years old and faced the possibility of a dire existence in Bavaria. But now there was an opportunity for an enterprising young man to travel to the United States where he would have the care and companionship of family. Heinrich was also her youngest son, and as noted, one she doted on and looked at as her favorite. However hard her decision may have been, she decided to allow Heinrich to go with his sisters to America. In her application to the Rossbrunn mayor for emigration, Margaretta wrote, "My minor son, Heinrich Schmitt, 14 years old, wants to emigrate to North America with his two older sisters. There lives an older daughter in America (Peru, free state, Ohio) past four years who is in very good circumstances. My son has only 200 Gulden as a bequest from his father. My daughter writes that there is a good future for all three. I beg now you will help me secure the necessary papers." With a further testament to his enterprising nature from his machinist employer, Heinrich was granted permission to leave Rossbrunn and sail to the United States. The loss of her family, so soon after the death of her husband, must have been devastating for Margaretta Schmitt. She died in Rossbrunn in 1857, and Heinrich never saw his mother again.[16]

Fourteen-year-old Heinrich Schmitt left Rossbrunn with his two sisters in May 1851, eventually reuniting with another sister, Anna. They were in effect joining a growing number of German immigrants to North America, a time during which the majority of German immigrants intended to do just what the Schmitt siblings hoped would be their fate: exchange the land hunger of home for a farm in the American Midwest or Southwest. In the United States, German immigrants tended to congregate in almost wholly German small cities and towns or in German neighborhoods in the larger urban centers. In this respect, German newcomers were no different from other immigrants to America. So many Germans moved into the American Midwest, particularly western Pennsylvania, Ohio, and Wisconsin, that the area became known as the "German triangle." In the cities and towns, the new Americans cultivated a specifically German culture, establishing their own press, choirs, theaters, gymnastics and shooting clubs, philanthropic organizations, and, of course, beer gardens. It was more a transplanting than an uprooting that was strengthened by the effects of "chain-migration," which forged links between particular areas in Germany and particular parts of the United States. The people of this German Diaspora remained strikingly unassimilated through the nineteenth century. Indeed, unlike her brother, Anna Schmitt would never truly learn English. Only in the following decades, and quite slowly, did the ethnic networks of countless "Little Germanys" lose their uniqueness.[17]

In the rural areas of the Midwest, chain migration led to the development of communities that reflected the origins of the settlers, so much so that even today German American communities retain traces of their founders' homelands through customs, dialects, and place names. In many cases, the aim was for German to be the official language, as ministers and teachers preached and taught in German. Peru, Ohio, was just such a place. Peru was located in Huron County, about fifty miles southeast of Cleveland. Germans made up the majority of the foreign migration to the county and, beginning in the late 1820s, established distinct German settlements in the southern part of the county. Anna Schmitt and her family arrived in Peru in 1848, settling where they found many former Bavarians in the community. Heinrich and his sisters arrived in Peru in late 1851. The Schmitts who remained in Bavaria learned of the successful journey to America of the rest of the family. In the Smith Papers is a letter received in Peru in 1857 from Johann, who, as indicated, had decided against traveling to America. In the letter, Johann expressed gratitude for receiving news of the travelers' safe arrival in Ohio. He also informed his sisters that their brother Peter would be joining them soon. Another Schmitt brother, George, would also sail for the United States, but before reaching his family members he died of yellow fever in New Orleans.[18]

Northern Ohio was a good location for a successful farm during the 1850s. When the Schmitts arrived, in 1851, the state had long ceased to be a frontier region and was instead a settled region of small family-run farms. Cincinnati was the state's largest city, although the completion of the Ohio and Erie Canal in the 1830s had caused Cleveland to become a boomtown. The typical Ohioan of the 1840s and 1850s lived in the country in a house of logs and operated a farm of no more than 125 acres. The general farmer grew grain, concentrating usually on corn and wheat, crops that Anna Schmitt and her family grew on their Peru farm. Through state and federal initiatives in the 1820s and 1830s, canals began serving the agricultural communities, linking them with markets in the East through the Ohio and Erie Canal. With access to markets via the canals, the price of wheat grew from a low of twenty-five cents per bushel in 1825 to over a dollar in the 1850s. For the Schmitts, the opening of the canal changed their lives and offered them some degree of prosperity since the Huron County town of Milan became an important inland port.[19]

Anna had been in America long enough to understand the need for education in the new land. She wanted Heinrich to go to school. Although he was older than most of the schoolchildren in Peru, Heinrich enrolled in the local school. He was a good student and evidently learned to speak, read, and write English very quickly. But the youngest Schmitt yearned to make his own way in life and seek an opportunity. He had never been a farmer in Bavaria, so life as one held little appeal. Besides, he was also now a young man living in a home with his sisters. Anna had a family of her own and Heinrich very well may have seen himself in the way. So, after less than a year in school, he packed his belongings and sought a job as a sailor on a freighter on Lake Erie.[20]

Even in the 1850s, most of the freighters operating on Lake Erie were sailing schooners. The new canals had brought economic growth to the region, and port cities, such as Cleveland and Sandusky City, were booming, leading to many opportunities for a young man to find work. Railroads by the 1850s had connected other parts of Ohio to Lake Erie, which made commercial freighting a burgeoning industry. Through the canal system, freighters on the lake shipped flour, corn, pork, bacon, and whiskey, predominantly to New York and eastern markets. For the sailors, the work was hard and filled with danger. Sailors commonly spent as much as four to five months on the lake, making numerous journeys across Lake Erie to unload their cargo at a port on the Erie Canal. They then picked up another load of finished goods for shipment back to more stops on the lake and then started the process all over again. It was backbreaking work that required hardy, strong men. Most sailors on the lake during this time were young single men hoping to make a few dollars and then begin an opportunity elsewhere, but the pay was meager and the work dangerous. Lake Erie is subject to powerful and sudden storms that claimed the lives of many sailors on its waters. It was a perilous job but also one that fifteen-year-old Heinrich Schmitt was eager to take on.[21]

Schmitt's career as a sailor did not last long. Less than a year after beginning his new job, the freighter he worked on was wrecked and the crew let go; the young man found himself unemployed. Reluctantly, his first big move having come to a swift end (something that would happen a number of times in his life), he moved back to Peru to be with his relatives. But his desire to strike out on his own did not end with the demise of his sailing career. After perhaps only a few weeks back with his relatives, Heinrich Schmitt had a new plan for his life. He again packed his bags, said goodbye to his family, and prepared to head west. He would not see most of his family members ever again, as it would be almost sixty years before he returned to Ohio. He set out on a course that would make him a "westering" man.[22]

Heinrich Schmitt, in 1852, began a journey that would change him from a young German immigrant into an assimilated American man. The first change he made after leaving Peru for the West was his name. Seeking an Americanized version of his family name, Heinrich became Henry Smith. He also must have felt that all Americans had a middle name so he settled on Clay. Perhaps he chose that name due to the popularity of Senator Henry Clay, a hero to many in the Midwest. For good measure, either personally or perhaps from a fellow traveler, he also received a nickname, "Hank." For whatever reason, Heinrich Schmitt was now Henry Clay "Hank" Smith.[23]

Although he was only sixteen years old, Hank Smith seemed to have no qualms about striking out alone. He never made his motives clear as to why he chose to leave the relative security of his sister's home, but that was probably because the reasons, to him, were obvious. After his father's death he had known poverty in Rossbrunn. He had also made the difficult decision to leave his mother in Bavaria and travel to a strange land. He might have chosen such a direction for one reason—his new home was the best place for him to find economic opportunity. He had first believed that a career as a sailor could help him attain his goals, but that scenario had sunk with his ship. When the prospects in Ohio did not seem clear, he saw no reason not to seek opportunity in the West. Young Hank Smith, although certainly courageous, was not seeking a life of adventure and romance. He was seeking wealth.

Sometime in late 1853 or early 1854, Hank Smith arrived in Westport (now part of Kansas City), Missouri. He made no note of how he traveled to Westport, how long it took, or if he worked at any point along the way. Westport was a popular destination in the 1850s, and there was a well-marked trail out of Illinois that made it a fairly easy trip. Westport was one of the primary "jumping off" points for trails west, including the Santa Fe Trail. From the time the trail was marked and laid out in the 1820s, the Santa Fe Trail was a lucrative trade route between the United States and the Mexican outpost of Santa Fe. But it was not an emigrant trail like the Oregon Trail. It was primarily a commercial road, as freighters carried thousands of tons of merchandise to a waiting market in Santa Fe. It was also often a two-way road and a meeting of cultures as Hispanic traders made journeys east to Missouri to enjoy the benefits of American markets and capitalism. By the time Hank Smith reached Westport, the Santa Fe Trail was a road between two American provinces and one reaping the benefits of more open trade and the increasing population in the Southwest. American and Mexican traders and merchants prospered by taking caravans of merchandise between Missouri and New Mexico, with many continuing to Chihuahua, Mexico.[24]

Some early scholars argued that the Santa Fe trade declined in the 1850s after the Mexican-American War. But Susan Calafate Boyle has made a convincing argument that the Santa Fe trade actually increased after the war, climaxing in the 1870s and dying with the arrival of the railroad. Primarily, she bases her contentions on the fact that most scholars have examined the trade only from the Anglo view and not through a study of the Hispanic New Mexican merchants and traders who played a large role in the lively trade. Through a careful examination of the records of the New Mexico merchants of the 1850s, Boyle shows that the amount of trade still going up and down the trail was significant and, in fact, higher than the levels of the 1830s and 1840s. For example, according to census data from 1860, the Otero family, father Vicente and sons Antonio José, Manuel José, and Miguel Antonio, were among the richest in New Mexico. Another Santa Fean, José Perea, amassed assets worth over $800,000, with the bulk of it accumulated during the 1850s. The point is that Hank Smith did not find a trade route in decline but one that was still dynamic and prosperous.[25]

Newly arrived in Westport, Hank Smith needed to find work. With the Santa Fe trade so prosperous, there were numerous jobs with a trading party. He signed on and became a teamster for the wagon master. The wagon master was the key figure in the whole freighting and trading operation. He had full responsibility for $18,000 to $30,000 worth of wagons, livestock, and accessories that belonged to someone else, unless he was a small operator who served as both wagon master and operator (as Smith would do later in his life). He was also responsible for $25,000 to $250,000 worth of goods that also did not belong to him. He supervised the loading of the wagons, making sure to balance the wagons' weight and bulk. A sample shipment might be tea, coffee, rice, sugar, tobacco, soup, candles, mustard, spices, and clothing articles. The wagon master, or an accomplished teamster, also had to have farrier skills in order to shoe oxen and mules, while also having enough mechanical skill to be able to repair wagons with the simplest of tools. Smith's trade in Bavaria as a machinist probably made him invaluable in this capacity. He also had to know where there was water and grass and where to halt and make night camp. He might be called on to hunt for fresh game to provide a meal that was a relief from the usual rations of sowbelly. Above all else, a good wagon master had to be able to get the best from both his animals and his men.[26]

A good wagon master was always in high demand, and operators took great care in selecting a man to lead their trains since the man had such great responsibilities and had to be inured to the difficulties of plains freighting. Nine out of ten started as teamsters and were advanced for faithful service. By the very nature of their job, a wagon master had to be strong, tireless, and, above all else, a leader of men since he had to exact obedience to his commands. Wagon masters on the Santa Fe Trail demanded and received high wages. While teamsters were drawing twenty-five dollars a month, the usual wage for a wagon master was at least one hundred dollars each month, good wages indeed in the 1850s. But just because their pay was high did not always mean they were pillars of the community. Although many wagon masters were of high moral character, an equal number of others were profane, hard drinking, aggressive men who often engaged in fisticuffs and, less commonly, gunplay. Young Hank Smith absorbed these lessons and, being a quick learner, decided to become a wagon master.[27]

Until the railroads connected the continent and penetrated into the interior of the nation, the only sure way to move bulk supplies, trade goods, and even heavy machinery was by wagon. Oxen predominantly pulled the wagons on the Santa Fe Trail, although hardy Missouri mules also made the trip. The freighters established a pattern of life for the men who worked the trade. In late February or early March, parties of men were sent out onto the plains to round up the oxen that had been wintering there, some as far away as Denver. The men drove the animals to the trailhead, and on approximately April 1 the first trains were loaded and rolled out, since starting at the beginning of spring meant that water and grass would be fairly plentiful. From Westport to Council Grove, the trip was comparatively easy, and then from Council Grove to the Great Bend of the Arkansas the trail crossed a pleasant country of rolling grassy plains, cut by many streams bordered with cottonwood trees. But west of the Great Bend there was a marked change in the appearance of the countryside: the green fields gave way to great reaches of short brown grass and cactus. Within a hundred miles of the Great Bend the train had to cross the Arkansas River. It was here the Santa Fe Trail split. One branch struck off southeastward across the panhandles of Oklahoma and Texas and into northeastern New Mexico. This was known as the Cimarron cutoff, or Cimarron branch, and was probably the route that Hank Smith took on his journey, since the other route, the Mountain branch, had experienced reduced traffic after the destruction of Bent's Fort in 1852.[28]

The Cimarron cutoff was a shorter route to Santa Fe, but it could also suffer from water shortages and the parties had to stay close to the foothills of the mountains where water and wood were available. Near the town of Watrous, New Mexico, the two branches rejoined. From there a single trail moved through Las Vegas, New Mexico, and then turned west and northwest through the Sangre de Cristo Mountains into Santa Fe.[29]

Teamsters on the trail were, by the very nature of their work, a diverse group. The productiveness of the Santa Fe trade meant that trains were frequent; crews could be hard to come by. Thus, wagon masters had neither the time nor the luxury to be picky or practice prejudice in selecting a squad. A frequent occurrence was a wagon train staffed by men of all different ethnicities and nations, most unable to communicate in a native tongue. Again, Smith was a quick learner and his skills probably helped him in getting a job.

Although he was little more than a year removed from Bavaria, he could at least communicate fairly effectively in English. Most crews of teamsters were divided in three "messes" of ten men each. The number-one mess was made up chiefly of professional bullwhackers, all Americans usually and the most experienced. They tended to lead the train. The number-two mess was mostly mechanics, who had been up the trail a few times and could be counted on to deal with most obstacles. The number-three mess consisted of the newcomers. They were usually immigrants or brand new teamsters fresh from Midwest farms (of which Hank Smith was both). Given the hierarchy of work, they had the least desirable jobs but also had the most to learn. Some stayed and became professional teamsters; others found the work distasteful and drifted off to other occupations.[30]

Hank Smith, probably without realizing it, had found his first career in the West. In his memoirs, he records little of his first trip on the Santa Fe Trail, so one must assume that it was routine and uneventful. However, given his later experiences as a freighter and operator, he must have learned much about the freight trade. Like many trains out of Missouri, Smith's did not unload everything at Santa Fe and return home. Santa Fe was a small city and could not absorb all the goods brought in from the trains. His train continued to push south along the Rio Grande, through the *jornada del muerto* into the Mesilla Valley and on to El Paso del Norte. Some trains continued from El Paso on to Chihuahua through the Chihuahua Trail, but Smith's instead chose to head west to Tucson with the purpose of unloading the remainder of its goods. The teamsters probably then rested a few days and then reloaded the wagons for the trip back to Missouri. Smith relates that the round-trip took the better part of a year, and that he was back in Westport by the spring of 1855.[31]

Perhaps Smith, arriving back in the spring, had missed finding work on another train headed for Santa Fe. Maybe he had had enough of the long Santa Fe Trail and a job on the third mess. Whatever the reason, he looked for another opportunity and found work with a surveying expedition. Congress had passed the Kansas-Nebraska Act in 1854, and a vital part of it had called for the organization of the Nebraska Territory. One of the first tasks after the organization of a territory was to make an accurate survey, a task usually handed over to the army; the new Nebraska Territory was no exception. A surveying crew gave Smith a new opportunity, so he joined one of the parties, under the command of Captain I. N. Palmer, and headed to Nebraska in late spring or summer 1855. The expedition was poorly organized and took longer than was planned because the party was still in the field when winter set in, forcing them to winter in the field. A snowy and cold winter forced the men to spend most of the season on Ponca Island in the middle of the Missouri River. They could not move until spring 1856, when they caught a boat down the Missouri River to Sioux City, Iowa. Sioux City was not much of a town in 1856, and the party eventually moved on to Council Bluffs in an attempt to get back to Kansas City.[32]

When the rest of the party trekked back east to Kansas City, Smith chose instead to try his luck in Iowa. He secured a job driving a team of horses, but the pay was poor so he moved across the river to take a job with the Council Bluffs firm of Boony and Armstrong. The firm had a contract to construct Nebraska's new capital building, and Smith was assigned the task of making brick. This job also played out fairly quickly, probably due to the stoppage of work during the winter. But Hank still needed a job, so he took work cutting winter ice on the Missouri River.[33]

After the winter cutting ice, Smith again found himself out of a job. He started down the river searching for work in the many towns spread along the riverbanks. The river towns, as they developed, were all much alike. At the river's edge was a levee, sometimes macadamized for all-weather use. Between the river and the bluffs was the business district, running back for two or three blocks. Here the warehouses and stores of the outfitters and the forwarding and commission merchants took up much of the space. Behind the business district and up the gullies that gave access to the tops of the bluffs were the small stores, saloons, and dance halls. On top of the bluffs was the residential section. Beyond the residential section were the wagon parks and the corrals of the freighters who sent their wagons into town in small groups to load at the warehouses.[34]

The buildings in the river towns were constructed mainly of wood and were often subject to disastrous fires. In the spring, mud was a common problem; at times it was so bad that often freight wagons bogged down in the middle of town. Public accommodations were crude, with several beds to a room and several occupants to a bed. Sanitation was sporadic or not present at all, and disease and stench were rampant throughout the towns. A constant source of difficulty for the freighter was the fencing in of the prairie outside the towns, which forced the roads into the wet stream bottoms and over the rolling hills. Public action to assure good access roads was slow coming. More often than not, the residents of these towns were mostly single men, which sometimes made for a rough and rowdy experience. The most plentiful businesses were saloons; the patrons consumed large quantities of the cheap whiskey sold in the establishments. The river town residents were not known for their piety, and churches were often difficult to find. Such a situation surely shocked the young devout Catholic from Bavaria.[35]

Smith stopped briefly at Kearney City and eventually moved on to Nebraska City, forty miles down the river from Council Bluffs. He worked for a short time as a handyman for a preacher in that city but eventually found a job where he could put his experience as a teamster to work. He took a job herding oxen with a fledgling freight firm that would become one of the most famous freight companies in the West—Russell, Majors, and Waddell. He helped outfit a wagon train for Fort Laramie and eventually, probably due to his experience as a teamster on the Santa Fe Trail, was hired on as a driver for the trip. Hank was still a young man, barely twenty-one, but he already had six years' experience as a freighter and was the youngest teamster in the crew.[36]

A job with the Russell, Majors, and Waddell firm was another vital step in training for the young immigrant. Alexander Majors entered the very competitive freight business with a different plan than most of his competitors. The majority of freighters used mules for trips into the far West, but Majors decided that oxen would make a better pack animal; they could forage on the plains, while mules could not, and six yoke of oxen could pull a wagon loaded with six thousand pounds of freight. Majors formed a partnership with William Russell and W. B. Waddell in 1855, and within three years the firm operated thirty-five hundred wagons and employed four thousand men. Armed with a two-year contract to supply the military garrisons in the Southwest, the company monopolized the trade in the region. Using the extraordinary profits from this operation allowed the firm to expand into stagecoaching.[37]

Smith successfully made the journey to Fort Laramie, but again he chose not to accompany his original party back to its origin. As the pattern of his life will suggest, he believed that the next great opportunity was just over the horizon. Smith was still in Fort Laramie when another wagon train arrived needing an experienced teamster to help lead them up the Oregon Trail. Although Fort Laramie was as far west as he had ever been, Hank immediately seized the opportunity and hired on to lead the train. It was not the usual trading party; rather, it was a Mormon cart train making the trip to Salt Lake City. The Mormons traveled in a train of over a hundred two-wheeled carts; the unusual characteristic of these carts was that neither mules nor oxen pulled them. Instead, they relied on men and women, which was typical in the early westward migration of Mormons, who often had insufficient means to purchase teams to pull the load. One or two people were assigned to each cart at the beginning of the journey (usually Iowa City), and they made the entire trip, a distance of over thirteen hundred miles, over some of the most desolate and dry areas of the country.[38]

Such a choice would seem strange for a man who had successfully worked as a teamster since coming west from Ohio. Pulling carts through the deserts of southern Wyoming and Utah for people of little means would not seem to be an advance in opportunity. Smith probably took the job as a means to an end; it would help him get farther west, where he believed he could find greater economic prospects. It was not an easy trail, and he only remained with the group until it reached Fort Bridger, where he took a job more suited to his experience—driving an ox team to Provo, Utah. But he did not stay with that train until it reached its destination either, leaving it before it reached Provo and traveling on into Salt Lake City.[39]

Smith was in Salt Lake City for a month before he secured another trailing job with an ox train headed for San Bernardino, California. Again, it was probably simply another choice made so he could get farther west, and it turned out not to be the best choice he could have made. One member of the party had kidnapped a Mormon woman before they left Salt Lake City, and the group had to travel along the mountains to stay concealed from a party of Mormon men sent to apprehend the kidnapper and the woman. Traveling quickly, the group had no time to forage or hunt, and the food ran out before they reached California, although the Mormon chasers probably gave up the hunt before they had even left Utah.[40]

Eventually, they reached California, where the group stopped at a small trading post in the middle of Death Valley. Again, a member of the party caused a conflict when he kidnapped a Paiute woman. The men were once more on the run and had to trek through Cajon Pass in order to keep another group of pursuers at bay. Eventually, they reached San Bernardino, and Hank left the troop, probably thankfully. In San Bernardino, Smith found another unusual and interesting job with a saloonkeeper. His job was doing something, as far as anything indicates, he had never done before: riding a bucking horse. His employer, John Lemons, used the horse as some kind of "come-on" for his saloon.

Men would gather outside the saloon at a corral and wager on whether Hank could ride the wild horse. If Hank stayed on, the men paid double for their drinks. If he was thrown, the drinks were free. This was certainly not the long-term occupation he had come to California for, and he soon quit, but it at least served the purpose of teaching him how to break a horse. He used this experience to convince a horse rancher, George Bird, to give him a job on his ranch in the San Bernardino Mountains. Smith spent the winter of 1857–58 on the ranch before he found another potential venture to chase.[41]

If he had been chasing the next proverbial gold mine throughout his journey from Ohio, this time Hank Smith literally chased a gold mine. While working on the ranch, he got word of a prospecting party headed to the Gila River in southern Arizona. The men told Smith that there were reports of new gold fields on the Gila and invited him to accompany them. Thirty men set out for Arizona in the summer of 1858. They arrived at Fort Yuma, bought some supplies and equipment, and traveled to Gila City.[42]

Over five hundred miners were working the river when they arrived. The men made camp on the river and then went prospecting in the mountains, as the Gila was much too crowded to even begin panning. The miners found some gold dust, but not much else. For one thing, the work was hard and water was scarce, making panning difficult. Mules had to pack water in each day, and the men were working in the heat of the Arizona desert. Finally, Indians attacked their camp and ran off the pack mules and horses, leaving the men without transportation and animals for packing.[43]

Eventually, after a few months, what little gold they had found became even scarcer. Some members of the party wanted to travel to the Colorado River, near where the river emerged to form the Grand Canyon, but the trip was too long and treacherous. Eventually, the prospectors decided to move up the Gila to look for new gold sources. But they found no good prospects and moved back toward Gila City. Here, they struck a placer about fifty miles from their original camp, which they called the Goose Creek Placer after a large number of geese they found near the area. Although this looked promising, after a few days they realized that it was not a large deposit and abandoned the claim. Back at Gila City they prospected a bit more and continued to find a small quantity of gold and, according to Smith, a few small diamonds. They next decided to sink a shaft to the bedrock in search of gold under the topsoil. Although they dug to a depth of two hundred feet, the venture proved worthless. The group of initially optimistic prospectors began to doubt their choice, and the group split up, some going back to California and others, including Hank Smith, traveling east.[44]

When the potential gold millionaires left Arizona, most of the party headed back for California. Smith had been to California and, despite what many people believed, did not feel he had any opportunities in the Golden State. Instead, the twenty-two-year-old Smith struck out east from Arizona, headed to El Paso, an area that would become his base and home for the next decade.

In the less than ten years since he left Ohio, Hank Smith had wandered all across the West, traveling in excess of two thousand miles. While it sounds adventurous, he was chasing something more tangible than adventure. He was chasing wealth. In all of his jobs, all of his travels, Smith was setting a pattern that would continue for the rest of his life; he was an entrepreneur, always searching for the next horizon that would make him rich. In El Paso and southern New Mexico he continued chasing that dream.

3

Seeking Opportunity In
New Mexico and Arizona

Hank Smith in a very short time had made his way to California, but mining for gold did not turn out to be as profitable as he had hoped. So, in Arizona, after his gold mining partners decided to return to California, Smith found himself in an increasingly familiar position—alone and in need of some means of support. After the theft of his horses near the Gila River, Smith traded his saddles for food ("two pounds of rancid mess pork, ten loaves of bread, four pounds of coffee, and five pounds of *pinole*") and began a walking trek over the Overland Mail Route, headed for the Rio Grande and El Paso. The Overland Mail Route in 1859 was a relatively new road. It followed the same course a US Army expedition had laid out in 1846 during the Mormon Campaign. By 1849–50, it had become a popular route for migrants to the California gold fields. Eventually, the road acquired the name the Southern Emigrant Road. It was one of the primary routes for a transcontinental railroad, and it became one of the reasons for the acquisition of the Gadsden Purchase.[1]

Broke and afoot, Smith traveled up the Gila River for 130 miles before stopping and finding work with the Overland Mail Company, which had begun operations on September 16, 1858, and was the first dependable transcontinental stage line. With the influx of migrants during the California gold rush and increased settlement in the western regions of the country after 1850, Congress decided, in 1857, that the country needed a western overland mail service to supplement the existing twice-monthly sea mail route between New

York and California. A central route through the heart of the country was probably the safest and most efficient, but antebellum political intrigue delayed the process. The postmaster general, Aaron V. Brown, a southerner, rejected any routes that did not travel through the South, hoping that federal contracts would attract greater capital to the South and further entrench the power of the southern slavocracy. Brown chose a route, which northerners called the "oxbow route," because of the deep swing the road took from the eastern terminals into the south, that ran from St. Louis and Memphis, through Texas, and across New Mexico and Arizona before reaching California.[2]

Congress authorized a $600,000 subsidy for the mail service but also mandated that the service would commence within one year. Twelve months would not have been enough time for some promoters, but for John Butterfield it was sufficient. Butterfield proved to be a good example of how far pluck and determination could carry an American in the nineteenth century. Born near Albany, New York, in 1801, Butterfield received little formal education. After finding a job as a stage driver at a young age, he rose to a controlling position in the stage firm and eventually dominated most of the passenger and mail stage lines in northern and western New York. Butterfield saw the potential of an "express" service in the 1840s that could safely and efficiently handle packages and merchandise as well as the mail. He founded an express service in 1849 and merged it with two other leading freighters of the day—Wells and Company and Livingston and Fargo—to form the American Express Company.[3]

When the chance came in 1857 to add the proposed overland mail contract to their other concerns, Butterfield and American Express pooled their resources with other eastern firms, formed a joint-stock association known as the Overland Mail Company, and won the contract. Butterfield became the primary organizer and promoter of the new company. To service the new route, he bought eighteen hundred head of horses and mules and over 250 Concord and Troy coaches. More than two thousand employees staffed the route, and the company expedited the building of stations across it. Butterfield spared no expense in getting the new company running, racing to meet the twelve-month deadline. He eventually overspent his budget and, citing the costs of his operation, the directors of Overland Mail forced him out of the company and into retirement.[4]

Smith arrived at the proposed Gila Ranch station at an opportune time. The construction of stage stations across Arizona was in full swing. The company plans called for stations approximately every twenty miles so that fresh horses and mules could be provided for the wagons and to give passengers accommodations and meals. The Gila Ranch station (referred to as Gila Bend on some maps) was roughly halfway between Tucson and Fort Yuma. After initially helping to build the station, Smith remained at Gila Ranch driving a hay wagon for the mostly Mexican forage crew. Smith and another man, Jack Pennington, drove the forage crew up into the lush spring-fed valleys in the higher elevations surrounding the desert plains of the station, where the Mexican laborers cut forage and loaded the wagons. Smith and Pennington then drove the wagons back to the station and stored the vital hay, along with corn, as feed for the station's horse and mule teams.[5]

For the Apache residents of Arizona, the Overland Mail service, and the establishment of the numerous stations, represented an unwelcome white encroachment into their lands. Smith received wages of eighty dollars a month (a high salary for the job and time) for his job as a wagon driver primarily because of the supposed Indian threat. The Apache peoples had first arrived in what would become the American Southwest at least by the early to mid-1600s, having apparently migrated to the southern Arizona region because of its proximity to Mexican settlements in Sonora and the possibilities for raiding there. Over the two centuries before Americans appeared in the area, the Chiricahua band of Apaches came to depend more and more on raids for livestock, including horses, which they used for food as well as transportation. By the early nineteenth century, most Chiricahuas had come to regard the people of Sonora not only as a raiding source but also as enemies, and a virtual state of war existed between the Chiricahuas and Sonorans.[6]

In 1850, a small party of Americans traveled through Chiricahua territory and informed the people that the United States had defeated Mexico in a war. For the most part, the Chiricahua considered the Americans to be allies who opposed a common enemy, but difficult relations soon developed. Most of the tension arose from disagreements about land and territory. Part of the problem was that neither Mexico nor the previous Spanish government ever had succeeded in occupying southern Arizona. When American soldiers told the Apaches that the United States now owned the territory, the position seemed, at least to the Chiricahua, absurd. The Americans told them that since the war had ended Apache raiding had to stop.[7]

American insistence had little effect on Apache raiding into Mexico, but during the early 1850s, little violence was directed at Americans. The Chiricahua leader, Cochise, then negotiated with the government in 1857 to allow Overland Mail carriers to cross Chiricahua territory unmolested. In some cases, the Chiricahua allowed outsiders to settle within their territory in exchange for payment. The Apaches almost certainly viewed this as a courtesy extended to allies rather than as a simple impersonal sale of land. For Apaches, the reciprocal exchange of gifts and hospitality was an important social interaction. But Americans often interpreted such encounters as transfers of exclusive rights to property. Still, the 1850s saw relatively little conflict between Native Americans and white settlers in Arizona. The principal reason was that, with the exception of a few mining ventures, the Overland Mail route, and some small ranches along the Santa Cruz Valley, there were simply too few white settlers around to raid.[8]

Smith reported only one Indian incident while he was at Gila Ranch, and it had little to do with the operation of the station. Smith and Jack Pennington heard about a herd of supposedly wild cattle near the mouth of the Salt River north of Gila Ranch. Not having any horses at the station at the time, the two men, accompanied by the station manager Abe Sutton, struck out on foot to round up the cattle, presumably to supply the station with fresh beef. After finding the herd, the three men butchered four of the animals and loaded the beef onto a small raft so they could float it down the Salt (a tributary of the Gila) and back to the station. The raft only had room for two men, so Sutton and Pennington manned the raft while Smith followed on foot on the bank. After going about a mile, a small Indian party fired on the raft, wounding Sutton in the wrist. Both Sutton and Pennington went into the river, and although Smith arrived and fired on the Indians after the attack started, the group lost most of the beef and the raft and had to walk back to the station. Smith did report some incidents of Indians stealing horses and mules from the station but, given the number of teams and men who passed through the station, even these incidents seem isolated.[9]

After the station was built and supplied, Smith left the Overland Mail Company and went to work for Abe Sutton, the station keeper, on Sutton's ranch. The ranch work would prove to be another in a long line of seminal experiences for young Hank Smith; working on Sutton's cattle ranch would be the first time Smith herded cattle, an occupation that he would return to almost thirty years later on the South Plains of Texas. Smith's main duty was to protect the cattle from predators and Indian theft. He was responsible for penning the cattle at night and guarding the cattle during daylight grazing. While he did tend Sutton's cattle on horseback, Smith's technique did not in any way resemble the Mexican-Spanish vaquero tradition. Smith dismounted and herded the cattle into pens at night and spent his days mostly watching the herd from the ground, mounting his horse only when a need arose. While he did once participate in a cattle "drive," it was a drive of only seventy-five miles to deliver some of Sutton's herd to his brother-in-law's ranch south of Tucson. Smith stayed at Sutton's ranch for only a few months before he moved on again, this time as a guard for the customs patrol, another job he would keep only for a few months. Eventually he found work again driving cattle to Mexico.[10]

Smith was setting a pattern that would follow him most of his life; he drifted from job to job. He was still a young man, just twenty-three, but he seemed to be looking for something that would finally make him his fortune. David Dary, in *Entrepreneurs of the Old West*, called men such as Smith the "silent army," a vanguard of merchants, mountain men, fur traders, freighters, homesteaders, and land speculators who gave the West a tradition of individual initiative in the pursuit of profit. While he may exaggerate the spirit of these western entrepreneurs, Dary correctly gauges the zealousness of the men in their search for profits and riches. If Hank Smith's intention had been simply to provide for himself at a rudimentary comfort level, he could have continued working at any of the number of jobs he had taken since leaving Ohio. But Smith, like so many people lured by the West's promise of opportunity, wanted more—complete financial independence.[11]

Hank found employment tending twelve hundred head of cattle for Attorney General William Wadsworth on a ranch in the Huachuca Mountains. Other than perhaps working with oxen while he was on the western trails, there is no account in Smith's papers detailing that he had previously worked cattle, thus it must be assumed that this was his first experience. After the cattle had fed for two weeks, Smith was to drive them to Fort Bliss. While the cattle were in the valley, Smith had invited a group of ten Chiricahuas into the camp and fed them. When it came time to drive the herd to El Paso, Smith decided to hire the Indians as herders, promising to pay them with a few head of cattle. The herd reached Apache Pass, a spot on the trail to El Paso notorious for Apache raids, without incident, and Smith allowed the Indian group to cut out four head. Since the use of Indian herders had worked so well on the journey through Apache Pass, Smith decided to try it again as he approached Cow Springs, where he found another clan of Chiricahua Apaches, the Bedonkohe, under the leadership of Mangas Coloradas.[12]

According to Mangas Coloradas's biographer, Edwin R. Sweeney, he, more than any other Chiricahua Apache leader, wanted to live in peace with Americans. Sweeney describes him as an "outgoing and gregarious man who preferred to resolve matters diplomatically . . . who actually trusted and truly respected Americans." Mangas Coloradas did not seek war with the American invaders, but in his later years as white settlers continued to encroach on Chiricahua territory, occupy prime Apache croplands, devastate the earth with mines, and slaughter or drive out the vital game animals on which the Chiricahuas depended he began to regard them as enemies. Resigned to no other alternative, in late 1860 Mangas Coloradas fought back and went to war.[13]

Such a description of Mangas Coloradas fits well with the man Hank Smith encountered at Cow Springs. At this point in his life, the Chiricahua leader was near seventy and still hoping for peaceful relations with American migrants to Chiricahua lands. Although he had been in Arizona only a short period, Smith had become quite proficient in the Chiricahua Apache dialect, and he approached

Mangas Coloradas about supplying some of his young men to help herd the cattle to the next water hole, Warm Springs. The Chiricahua leader agreed to help Smith and invited Smith and his men to spend the night at the Indian camp. The next morning Smith hired ten Chiricahua men at a dollar each to help drive the herd and to serve as night sentries. The Chiricahuas scouted a good trail for Smith to Warm Springs and helped bring the herd to good water. The party decided to spend some time camping near the water and spent the next few days "feasting, resting, and running foot races."[14]

Smith finished his cattle drive in El Paso in either late 1859 or early 1860, and, although he had intended to remain in the El Paso area, caught gold fever once again. Hearing about the possibility of gold in the Pinos Altos Mountains in southwestern New Mexico, Smith embarked on the gold mining expedition with eleven Americans and three Mexicans to locate an old mine in the region. Before reaching the Pinos Altos, the men found the original placer, but after three weeks they decided that it was played out. They moved farther into the mountains, and in May 1860 three of Smith's companions, including veteran prospectors Jacob Snively and two men named Birch and Hicks, made the first strike of gold in the Pinos Altos Mountains. The miners originally called the site Birchville, but soon renamed it Pinos Altos for the tall ponderosa pines that were found in the region. News of the discovery spread, and a boomtown developed within days as prospectors swarmed into the mountains. According to Smith, within one month five hundred miners had arrived at Pinos Altos. Smith's account of the discovery closely parallels accounts in other historical works.[15]

The Pinos Altos gold fields proved to be quite profitable. Smith reported that the claims averaged from forty to fifty dollars a day per man, but expenses were also very high. Lumber was two hundred fifty dollars per thousand pounds of board feet, and water was fifty cents a gallon, if any water was available. Fresh meat was scarce and expensive, and during the winter supplies were at a premium. The gold rush town of Pinos Altos was also a typically crude and sometimes violent place. Drunkenness was a constant problem and disputes over claims a regular occurrence. Smith even engaged in a "duel" with a man named Billy Estelle, although presumably neither man was harmed since Estelle's name appears on the muster rolls of the Arizona Rangers in 1861.[16]

James Tevis, in an article in *True West Magazine*, includes a description of the new town. He recounts how shortly after the gold discovery there were "saloons, dance halls, and a tenpin alley ... with gamblers galore." Sunday was reserved for the settling of disputes, "unless contending persons were full of whiskey, in which case the killing was quickly done." Roy Bean (who would later become notorious as a "judge" in far west Texas) and his brother, Samuel, operated a mercantile store in Pinos Altos, an establishment that dealt in "merchandise, liquor, and a fine billiard table." At its height in 1861, Pinos Altos boasted a population of close to one thousand people.[17]

The discovery of gold and the gold fever boom that followed in the Pinos Altos region precipitated a turn in the relations between the Apaches and Americans, a consequence that would again bring Hank Smith into contact with his one-time friend Mangas Coloradas. This time the two men would not be allies or friends but on separate sides of a violent conflict. While Pinos Altos was filling up with new residents searching for riches, Mangas Coloradas remained at Santa Lucia Springs, planting some crops and still avoiding contact with Americans. But the Chihennes band (to whom Mangas Coloradas was affiliated by marriage), whose territory had been overrun with the miners at Pinos Altos, was not as inclined toward peace as Mangas Coloradas and his forty-seven-member Bedonkohe band. Raiding into southern New Mexico, principally the city of Mesilla, became more common, and the new American settlers often retaliated. One of the most infamous retaliatory incursions occurred in December 1860. A group of thirty well-armed miners, most of them from Texas, ambushed an Indian camp on the west bank of the Mimbres River, killing four Apaches. James Tevis, the group's leader, claimed that the encamped Apaches were part of a group that had stolen horses at Pinos Altos, although the evidence suggests that another Indian group was probably responsible. As Apache peoples began to confront their American invaders, Mangas Coloradas was forced to abandon his notions of peaceful relations with the Americans, and an Indian war ensued in eastern Arizona and western New Mexico.[18]

According to Edwin R. Sweeney, "in . . . 1860 the last thing Mangas Coloradas had on his mind was going to war with Americans." Granted, with the appearance of Americans in Pinos Altos and the raids into New Mexico, relations were deteriorating and increasingly becoming a source of problems. The incursion of miners into the heart of Mangas Coloradas's region had destroyed Apache lands and psychologically devastated the Indians. The invasion of Americans, and the destruction of the land and game that often followed, appalled the Apaches, for they had never witnessed such an influx of aggressive, profit-driven, Americans into their lands. The new Americans were not the trappers and mountain men the Apaches had encountered before, men who often sought friendly relations with Indian peoples, or the travelers simply moving farther west. Instead, these Americans were frontiersmen, many from Texas, usually well-armed and openly contemptuous of Native Americans. A sense of despair must have gripped the once powerful bands of Chihennes and Bedonkohes, since just a decade before they would have intimidated or evicted the white invaders.[19]

According to Chiricahua history, Mangas Coloradas occasionally went to Pinos Altos, alone or with a few of his people, to trade or perhaps to affirm his desire for peace. Such behavior would fit with his character and personality, a desire to try to make peace, especially with Americans, whom he had somehow come to respect. During one occasion, Mangas Coloradas was apparently the victim of a brutal and unprovoked attack, one that Smith mentions in his accounts of Pinos Altos. It is confirmed through other accounts, including one by John Cremony, a Boston journalist, but Edwin R. Sweeney calls the story apocryphal and a legend. However, Smith claimed to be present during the incident and his account closely parallels Cremony's. As the story goes, Mangas Coloradas approached the miners of Pinos Altos and promised to show each man another place where gold was "far more abundant and could be attained with less labor." For the Chiricahua leader, this may have been what he thought was the best way to get the white men to leave. Smith recounts that the next time Mangas Coloradas appeared in the camp, a group of miners grabbed him, tied him, and whipped him with a "blacksnake whip." Smith's account records that "Mangus [sic] Colorado [sic] neither winced or uttered a sound while the lash bit deeper and deeper into his flesh, cutting the skin on his back to ghastly ribbons, bringing the blood in streaming rivulets to trickle down his legs."[20]

Did the alleged beating truly take place? Sweeney doubts the veracity of the story, citing that John Cremony was prone to exaggeration and his stories should be considered unreliable unless confirmed by corroborating accounts. But Hank Smith recounted the same incident and claimed to be present during the beating, a recollection that Sweeney claims is not present in the Smith memoir. But it is, and according to his sources, Sweeney examined Smith's papers. He also doubts the beating story since no other contemporary accounts mention the incident and Chiricahua history also does not record the event.[21]

Furthermore, Sweeney doubts the incident truly occurred since Cremony describes the event as taking place in December 1860. According to Sweeney, Mangas Coloradas had already declared war on the Americans by that time and clearly would not have entered the mining camp. Smith's recollection gives no date, and his memoirs are very difficult to interpret beyond approximate dates.[22]

Sweeney conceded that the episode could have happened earlier than December, but he discounts it because of contemporary events that, in his mind, preclude Mangas Coloradas visiting the camp. But, Mangas Coloradas himself cryptically mentions the event when he said in 1862 that "he was at peace with the world until the troops attacked and killed many of my people," and after the third assault he "armed himself in self defense." The first two assaults are clear: Tevis's attack in August 1860 and an attack by a group of white settlers on Cochise at Apache Pass. The third attack he mentions could quite possibly be the incident at Pinos Altos. The attack in Pinos Altos would also explain the enmity Mangas Coloradas felt toward Americans in the last years of his life. Whether or not the beating occurred remains ambiguous.[23]

Although the alleged beating of Mangas Coloradas is dubious, war between the Apaches and the residents of Arizona and New Mexico in 1861–62 is not. The Chiricahuas, angered at the incursions on their land, the attacks against their camps, and the incident at Apache Pass, made the decision to go to war. Mangas Coloradas, who had advocated peace with the Americans for most of his life, probably felt that he had no choice in the matter. By early 1861 his son-in-law, Cochise, had launched his self-prescribed mission of driving Americans from southern Arizona, a decision that coincided with the white's virtual abandonment of the territory. White settlers in both southern Arizona and New Mexico fled to the east, and stations, forts, and ranches were quickly abandoned. For the Chiricahuas, such abandonment had to be the result of Cochise's efforts, and he claimed responsibility.[24]

In summer 1861, the two Chiricahua leaders joined forces and established a base near Dziltanatal (which white settlers called Cooke's Peak) in Luna County, New Mexico, about thirty miles from the present-day city of Deming. As the Americans abandoned the area, the Chiricahuas ambushed and killed as many white settlers as possible. One estimate claimed that between 1861 and 1863 over one hundred white settlers were killed in Cooke's Canyon, a report that, unlike so many other reports concerning conflicts with Indians, appears not to be exaggerated. The continuing conflict further isolated the miners at Pinos Altos, curtailing mining activities and subjecting the miners to livestock raids; essentially, the Pinos Altos residents were under siege and could not move out of the immediate area of the settlement.[25]

Because of the Chiricahua uprising, the miners and residents of Pinos Altos had to either leave their claims or organize some sort of resistance. In 1861, the miners organized the Arizona Guards (the name presumably comes from the fact that the miners were "guarding" against Indian attacks from Arizona) under Captain Thomas J. Mastin. Although records indicate no earlier formal organization in the mining town, Smith's version of events suggests that there was some sort of local militia under the command of Thomas Helm in 1860. Mastin had chosen Thomas Helm to serve as a lieutenant of the Arizona Guards, and he became the commander after Mastin was killed. Smith never refers to Mastin as commander, preferring to name Helm as the leader of the group during his entire tenure. The newly formed Arizona Guards would soon be involved in a large-scale battle with the forces of Cochise and Mangas Coloradas.[26]

The fight began at the Chiricahua headquarters at Dziltanatal when the Apache forces attacked a group of American ranchers fleeing Arizona for Mesilla. Known as the Ake party (named for Felix

Grundy Ake, a fifty-year-old farmer), the wagon train consisted of six double wagons, two buggies, and one single wagon. The men also carried their possessions and herded several head of cattle, sheep, goats, and horses. Such a bounty made them an inviting target for the Chiricahua.[27]

Ake had intended to meet a group of soldiers who had departed Fort Buchanan after he left Tucson, but the soldiers had already left. Ake had no idea that Cochise and Mangas Coloradas had combined their forces, so, believing he had enough men to repel any attack, decided to continue on to the Mimbres River. Ake and his party then camped at an abandoned stage station. While in camp they heard a report that a party of over two hundred Apaches had annihilated nine Mexicans at Cooke's Canyon, but Ake discounted the story. He should have heeded the warning.[28]

Hank Smith offered a contemporary account of the doomed Mexican party. Anton Brewer, a butcher at Pinos Altos, had gone to Mesilla to buy some cattle for the starving citizens of the mining town. He bought forty head of cattle and employed nine Mexicans to herd the cattle to Pinos Altos. Although they knew it was dangerous, they decided to return through Cooke's Canyon, a site of recent Indian attacks. Smith (who obviously heard the story from Brewer) reported that the Apaches quickly surrounded the cattle drovers and killed them all. Brewer, who was guarding the cattle away from the camp as his men ate, heard the attack and immediately rode away, reaching Pinos Altos with the news. Smith was one of fifteen men who went to Cooke's Canyon to bury the bodies. He stated, "they were found in a close pile, horribly mutilated." Smith's recounts of mutilation could be an exaggeration. In previous engagements both Cochise and Mangas Coloradas showed admiration for the bravery of their enemies and forbade any mutilations. However, since these men were Mexicans, the sworn enemies of the Chiricahua, the two Apache leaders may have reacted differently.[29]

The Ake party, dismissing the report as rumor, left the Mimbres and made their way toward Cooke's Canyon and the pass to Mesilla. Cochise and Mangas Coloradas, with close to two hundred men, prepared an ambush in a narrow part of the canyon, not too distant from the spot where Brewer's men had met their fate. About midmorning, the group's stock entered the canyon first, followed by the wagons and a few stragglers. One of the men soon discovered that the report about the cattle herders was true. Tripping over the corpse of one of the vaqueros, he reported his find to Ake, and the men gathered around the entrance of the narrow gap. Ake was still not convinced about the danger and ordered the stock and wagons ahead. Waiting until the canyon walls hemmed in the entire group, Cochise and Mangas Coloradas ordered their men to attack. They instantly killed two men as the rest of the party formed a breastwork within a group of wagons. The Indians rounded up the stock and drove it to the opposite end of the canyon, and the battle gradually evolved into a sniping duel between the two adversaries. But Cochise and Mangas Coloradas had achieved their objective—they had the stock and had killed a few Americans. They felt there was no reason to risk additional casualties and withdrew. Having suffered a loss of four dead and between five and eight men wounded, the Ake party limped back to the Mimbres.[30]

After reaching the Mimbres stage station, Ake sent a messenger to Pinos Altos to ask for assistance from the Arizona Guards. According to Smith, the guards waited until the next day to move toward the stranded Ake party, but other reports remember the guards meeting Ake's survivors as they reached the Mimbres, at least a two- or three-day ride from Pinos Altos. If that is correct, the guards must have mustered in response to the news about Brewer's party and narrowly missed encountering the Chiricahuas as the Apaches attacked Ake's group. Smith's account of Indian fights in the Pinos Altos region also has another discrepancy. Smith dates the encounter with the Chiricahuas as occurring in September 1860, but all other recollections of the battles give the date as September 1861. While this may seem insignificant, if the true date of the fighting is 1861, then it occurred after the Arizona Guards had been mustered into the Confederate Army in New Mexico under Lieutenant Colonel John Baylor. Baylor charged the guards with the duty of fighting the Apaches and "to reopen the road between Tucson and Mesilla." Such a scenario would mean that the guards rode out not as a revenge force but one with at least an authoritative charge to guard travelers against Apache attacks.[31]

Many of the men knew the Chiricahuas from their trading expeditions at Pinos Altos. Mastin knew Mangas Coloradas and had a great deal of respect for him, declaring that he was the Apaches' "ablest statesman." Of course Smith and Mangas Coloradas were also acquainted. Whatever their opinions of the Chiricahuas, Mastin's Arizona Guards met Ake's survivors and mounted an expedition to catch the Apaches before they made the safety of Mexico. Mastin decided to ride straight for the Florida Mountains, hoping to get ahead of the Indians and engage them while they were traveling slowly with the captured stock.[32]

After riding all night, the Arizona Guards had indeed gotten ahead of the Apache party and prepared an ambush. As they watched the Apaches approach, the hidden guards no doubt could scarcely believe their luck; the Chiricahuas were riding right into an ambush not unlike the one they had performed on Ake's party. When they got within a quarter of a mile, the Arizona Guards leaped out into the open and charged the surprised and fatigued Indians. Hank Smith reported that he was one of the first to "run into" the Apache party and that the white men killed eight Indians in the initial assault. Smith and his compatriots pursued the Indians across the open plains until the Chiricahuas found refuge in the foothills.[33]

More than likely Smith and the Arizona Guards had struck Cochise's group. When Mangas Coloradas's forces made their way to the mountains, the Indians must have decided that they could not let the white fighters' response go unanswered. The two Indian leaders very well may have been completely astonished that the Pinos Altos men so brazenly attacked their war party. Just a short month before they had watched their new enemies desert their posts and move eastward, presumably in response to the Chiricahuas' open hostility to their presence.[34]

But, Mangas Coloradas and Cochise had no way of knowing that the withdrawal of the US troops had another reason—the beginning of the Civil War. The US Army's departmental headquarters in Santa Fe had ordered the troops at Fort Fillmore to move east and prepare to stop the anticipated Confederate advance into New Mexico. Northern leaders knew that Confederate States of America president Jefferson Davis not only coveted the resources of the Southwest and

California but also desperately wanted a sea outlet in southern California as a means to continue trade and counter the Union blockade. A Confederate detachment, under Lieutenant Colonel John S. Baylor, occupied Fort Bliss in Texas after federal troops had evacuated and prepared to move into New Mexico. Baylor's forces marched into New Mexico and captured a federal detachment near the Organ Mountains without firing a shot. After the disaster, the federal troops at Fort Stanton abandoned their post, leaving southern New Mexico at the mercy of Baylor and the Confederates. Baylor moved into Mesilla and immediately declared himself territorial governor and began calling for volunteers to repel federal troops from New Mexico and Arizona. In fact, by the time the Arizona Guards had attacked the Chiricahua in the Florida Mountains, Baylor had already mustered the group into the Confederate Army.[35]

While secession and war with the Union may have occupied the Pinos Altos men's minds, the more immediate crisis was the Chiricahua insurrection. In mid-September, Cochise and Mangas Coloradas led their men toward the settlement of Pinos Altos, hoping to rid their territory of white men for good. Their numbers grew, reinforced with other Chiricahua bands from Mexico. In the early morning hours of September 27, 1861, the Chiricahua leaders assembled their followers on the outskirts of Pinos Altos, which consisted of several mining camps. Cochise and Mangas Coloradas assigned a cluster of fighters, each under a local war leader, a particular camp to strike.[36]

Hank Smith and the Arizona Guards arrived in Pinos Altos before daybreak. Smith, after first stopping to secure provisions, began the trek to his camp at Birch's Gulch when he heard firing at least a mile away. Although the sound of gunfire was fairly common in the miner's camp, Smith immediately suspected an Indian attack. The Chiricahuas' initial assault caught the mining community completely by surprise. Cochise and Mangas Coloradas had formulated a carefully conceived plan, one that they had used numerous times in raiding Mexican towns. The assembled small bands simultaneously struck each camp, confusing the miners about the direction of the attack and delaying a defensive response. The battle swirled all around the town, and many miners were either unable or unwilling to come out and fight. Smith asserts that it was fright that kept many miners from responding, a development that almost led to complete capitulation of the town.[37]

Smith and three other men rushed inside his cabin and, under cover and with the fresh provisions Smith had bought in town (including ammunition), repulsed the Chiricahuas surrounding his home. After repelling the Apaches, Smith and the men rushed to the main camp. Inside the settlement of Pinos Altos, Smith described the fight as "hand to hand" and obviously a scene of great carnage. Early in the fight, the Chiricahuas had tried to burn the log houses, an attempt that for the most part failed. The fierce fighting continued through the morning, and by noon the battle was concentrated around the Bean store. During the fighting around the store, an Indian bullet killed Tom Mastin and the Chiricahuas appeared near victory. But, shortly after noon the miners, with the aid of several women who were caught in the middle of the battle in Bean's store, loaded a small cannon with nails and buckshot and fired the weapon directly into the Apache lines. This staggered the Indians and allowed the miners to make a furious counterattack, resulting in a rout that drove the Apaches from Pinos Altos.[38]

Casualties on both sides were high. Smith reported that the Apaches left ten dead on the battlefield, and he believed that they carried off twenty more dead and wounded. Of course, such a claim is pure speculation and one that many men made during fights with Indians. Smith's recollections were also recorded fifty years after the battle and in a different era. The Apaches killed five men, including the Guard's leader Tom Mastin, and seven other men suffered a varying degree of wounds. The miners claimed a victory, but it was a hollow one in most respects since, in the days that followed, many miners abandoned their diggings for Mesilla or Santa Fe (where they went largely depended on their views about the Civil War) until only seventy miners and a detachment of Arizona Guards remained at Pinos Altos.[39]

Hank Smith was one of the miners who decided that his fortunes lay outside Pinos Altos. He and some other miners decided to take a prospecting trip northwest of Pinos Altos. Smith's descriptions of the area suggest that the group passed over into present-day Arizona. The prospectors found no gold on the trip and eventually turned south, where one of their members found a gold strike near the San Francisco River. They worked the site until dark but found no other significant finds. Smith was convinced that gold was in the area and secured a claim on a parcel near the river. Eventually he returned to Pinos Altos, where he found the town still full of talk about war and secession. Most of the miners at Pinos Altos, which probably contained a majority of Texans, favored secession and the Confederacy, but Smith seemed to have no opinion one way or another. He was only twenty-five and had been in the United States for just nine years. He was a German immigrant who had spent most of his time in America wandering the West looking for his fortune. He had more than likely never given the whole issue of slavery, states' rights, or secession even minimal thought. But Hank Smith would soon be caught in the fervor and emotion of a Confederate invasion of New Mexico, and he would have the unique distinction of serving on both sides of the conflict.[40]

4

Hank Smith in Gray and Blue

Although Hank Smith and the Arizona Guards' primary preoccupation during 1861 was the Chiricahua, the specter of Civil War preoccupied their minds. When the Confederate States of America was formed, there was no doubting the southern resolve, but the men, arms, and wealth needed to combat the overwhelming economy and manpower of the United States was in short supply. But potential for men, arms, and most of all gold, lay in the territories of New Mexico and Arizona. Confederate leaders reasoned that if the CSA could take and hold the two southwestern territories they could secure the arms stored at the frontier forts, and the residents of the region would flock to the southern standard, giving the fledging nation soldiers, munitions, and access to the gold fields of the region. The ultimate dream for the Confederacy was to establish a southwestern empire that would stretch all the way to the Pacific Ocean, allowing the South access to the Pacific trade and an expansion of the dominant slavocracy.[1]

The Texas secessionists took the lead in securing control of Arizona and New Mexico for the Confederacy. During the Montgomery Convention that established the CSA, Texas delegate William Reed Scurry introduced a resolution to appoint commissioners to deal with the territorial governments on the Confederacy's behalf. The '62 Convention chose as one of the commissioners Simeon Hart, a well-known and respected leader in the El Paso area. A resident since 1822, Hart was a prominent trader on the Chihuahua Trail between Santa Fe and Mexico and had established a mill on the northern side of the Rio Grande across from El Paso del Norte. He enjoyed good relations with the residents of New Mexico and employed a large number of the Hispanic population in the region. Southerners and Texans anticipated that the West would join their secession, largely due to the fact that the residents of southern and western New Mexico were becoming increasingly frustrated with the federal government's seeming inaction and lack of progress in suppressing Indian raids. In fact, a southern New Mexico faction had taken advantage of the sectional turmoil in the rest of the nation and declared the territory of Arizona separate and apart from New Mexico. Since the US Congress had refused to recognize this action, the Confederates sensed an opening and sent their delegation to meet with the leaders of this rogue territory.[2]

The remainder of New Mexico also seemed to share sympathy and confederation with the South. Federal army officers in Albuquerque reported that New Mexicans seemed to favor secession. The territorial leadership tended to reinforce such sentiment, as over the previous four years the Buchanan administration had appointed far more southerners than northerners to territorial government posts. Southerners filled the two most important territorial positions: Territorial Secretary Alexander Jackson and Governor Abraham Rencher. The New Mexican territorial delegate to Congress, Miguel Otero, had also cultivated southern friends, strengthening the New Mexican tie to the South. Southern partisan residents of New Mexico had also succeeded in passing a slave code in the territory in 1859, even though the region had fewer than a dozen slaves in 1861. Finally, the editor of the influential *Santa Fe Gazette*, Samuel Yost, encouraged alignment with the slave states.[3]

The road to a Confederate southwestern empire would begin in Texas. Secession offered the state new avenues for expansion and the achievement of finally fulfilling the Texas dream of acquiring Santa Fe or at least establishing a trade relationship with the major New Mexican city. Confederate acquisition of New Mexico and Arizona would also help Texas achieve a satisfactory solution to the question of its western boundary, and Texan armies could, perhaps, use the excuse of war to move into and establish client states in Northern Mexico, with possibilities of future annexation. James Reily, who eventually became a Confederate colonel and served under General

Henry Hopkins Sibley in the New Mexico campaign, made clear the ultimate Texan goal when he stated, "we must have Sonora and Chihuahua. With Sonora and Chihuahua we gain California, and by a railroad to Guaymas render our state of Texas the great highway of nations."[4]

John Robert Baylor became the first man to lead a Confederate military force into the New Mexico territory, and Hank Smith eventually served under his command. Baylor was a notorious figure on the Texas frontier. As secession and civil war seemed to create an opportunity for Texas, Baylor also saw the crisis as a personal opportunity. Like Hank Smith, John Baylor came west from Ohio, and he also saw the West as the best place to gain fame and fortune. Unlike Hank Smith, however, his plans were much more grandiose. Baylor left Ohio as a fourteen-year-old boy in 1836 to join the seeming romance of the Texas revolution. Traveling with his older brother Henry to join another sibling already in Texas, the Baylor brothers only got as far as Kentucky, where a man they met in a tavern persuaded them to return to Cincinnati and continue their studies.[5]

Baylor did not abandon his Texas ideas. In 1840 he again left Ohio and settled in Fayette County, Texas, and, with the Texas revolution finished, had to settle for campaigns against the Comanche and expeditions of fortune and glory. In 1842, he joined a volunteer militia raised for the disastrous campaign against Mexican troops occupying San Antonio. Only a rampaging creek and the loss of his horse kept Baylor from the massacre of the Texan force at Salado Creek.[6]

Through the next few years Baylor attempted to follow a more stable path, becoming a teacher on a reservation in the Indian Territory, but he was once again embroiled in controversy when he was accused of murdering a white trader. He left the reservation with a price on his head and returned to Texas, settling in Marshall and marrying Emily Hanna. In 1845 he returned to Fayette County to try his hand at farming. For the next five years he found stability and started raising a family on his farm near La Grange.[7]

In 1851, Baylor decided to pursue politics, and the voters of Fayette County elected him to the state legislature. In Austin he cultivated political contacts; he became a lawyer in 1853. During his tenure in Austin, Texas established a Comanche reservation near

Fort Belknap on the Clear Fork of the Brazos River. He accepted a position as an Indian agent on the reservation and in 1855 moved his family to the Clear Fork. But his unorthodox approach to relations with the Comanches again led him to controversy.

He often accompanied the Indians on buffalo hunts and was a frequent visitor to their homes. He also allowed the Comanches to camp outside the reservation's borders; such laxity eventually led some reservation Comanches to join war parties in the Indian Territory. Eventually, reservation director Robert S. Neighbors determined that Baylor had lost all control of the reservation. The ensuing investigation led to accusations against Baylor for financial corruption on the reservation, which led to his dismissal in 1857.[8]

Baylor blamed the improprieties on the Comanches and not on his policies and actions. The former Indian agent then seemed to take out his anger on all Indians. After moving his family to Parker County, Baylor became one of the leading critics of reservations and led at least one attack on reservation residents. He also became the publisher of a racist newspaper, the *White Man*, which repeatedly called for the complete destruction of Native Americans. During this period, the former northerner also became committed to the idea of secession and joined the notorious Knights of the Golden Circle. Finally, in February 1861, he left his family again, this time for the cause of secession, and journeyed to San Antonio to join the Texas forces that were massing for the march to New Mexico.[9]

Henry Hopkins Sibley was placed in command of the forces at San Antonio. General Sibley was an ex-US Army officer who had a reputation as being brave and capable. But the recollection of William Davidson, a Texas soldier under his command, was that "no matter what he might have been in the old army, when he came to us he was the very last man on earth who ought to have been placed in command of that expedition." Sibley had held the rank of major in the federal army before resigning to join the Confederate cause. In the Confederate forces, he gained the rank of brigadier general and, according to numerous accounts, was so elated at his promotion, "he had not the prudence to keep his tongue still but gave it free rein and espoused all his plans so that the enemy generally knew as much about his plans and intentions as he did himself." The Confederate invasion force lingered in the San

Antonio area well into November and faced the prospect of marching across the barren west Texas plains in the dead of winter. Sibley is responsible for much of the delay, as he was tardy in arranging transportation and, according to Davidson, "spent much of the fall drunk."[10]

In advance of the invasion, the Confederate commander of the Department of Texas, Earl Van Dorn, ordered John Baylor and the six companies of the Second Texas Mounted Rifles to proceed to El Paso and occupy the now abandoned Fort Bliss. Once in El Paso, Baylor was to reconnoiter the area in southern New Mexico and, if possible, move into the territory and establish a position. Baylor viewed this as an opportunity to restore his military and personal reputation. By advancing to Fort Bliss and then into the Mesilla Valley, not only would he be a candidate for military glory he would also be leading a group of armed Texas men into a region that the state coveted and had once claimed. If the CSA could win the war, Texas stood a chance of extending its western border. Baylor also saw the move into New Mexico and Arizona as another chance to continue his personal war of hatred against Native Americans.[11]

Indications suggested that support for the Confederacy was strong in the New Mexico Territory, particularly from the citizens of Tucson, Pinos Altos, and Mesilla in the Gadsden Purchase region of the New Mexico territory. Hank Smith was caught in the middle of the war talk. Smith contradicts some other reports as he recalled that Pinos Altos was "evenly divided . . . some in favor of Uncle Sam and some in favor of Jeff Davis, most in favor of no war at all. The men [of] southern birth . . . got all hot-headed and made threats to those that were not inclined that way." Smith's recollections of the time may reflect more his personal feelings than they do the actual situation in Pinos Altos. Most of the men at Pinos Altos were southerners, the majority of those from Texas. It is safe to assume that the men would favor secession and war with the Union if it became a reality. But Hank Smith had spent no time in the South; he was a German émigré who had spent most of his time in the United States in the West. The southern cause of slavery, states' rights, and the concept of southern tradition and honor probably held little appeal for a man who had left Germany and Ohio with little more on his mind than finding economic opportunity. Although he never made his true ideas about slavery and the war known, Smith showed little enthusiasm for supporting the Confederacy, although he would eventually muster into the Confederate Army with the rest of the Arizona Guards. Perhaps his decision was as simple as being a man reluctant to abandon his friends. His decision may also have to do with the fact that the mines were all but abandoned by the summer of 1861. Some of the men had left to join the war; others had moved on to the gold fields of Colorado. Once again, Smith was left largely adrift and probably believed that joining the Confederate Army with the Arizona Guards was his most prudent course.[12]

Baylor and the Second Texas Mounted Rifles arrived in El Paso in June 1861. By July 1, they had secured Fort Bliss and with it at least a year's supply of rations and provisions. Baylor would use Bliss as a base of operations for the second phase of his plan—securing southern New Mexico (or Arizona as the Confederates referred to the region) for the Confederacy. Baylor and his men became the advance guard for the dream of a Confederate western empire.[13]

As Baylor and his men organized around Fort Bliss, US Army Colonel Edward R. S. Canby, the commander of the federal Department of New Mexico, prepared for the invasion. The US Army command had ordered all federal troops in New Mexico to abandon the territory and move east, so Canby called for volunteers and militia to muster to meet the threat. Five thousand New Mexican volunteers heeded his call, almost all of them from the northern portion of the state. He then evacuated most of the forts in southern New Mexico but concentrated a small force at Fort Fillmore, the closest installation to Fort Bliss. Baylor decided to seize the initiative before the federals could mount an effective defense and, under the cover of darkness, led 258 men toward Fort Fillmore on July 23. The Confederates appeared to gain the advantage as they positioned themselves between the post and the only available water supply. But Baylor was relying on the element of surprise, and that was spoiled when a Rebel deserter informed the Fillmore forces of Baylor's waiting trap. With his presence known, Baylor withdrew his men across the Rio Grande and entered Mesilla, whose citizens welcomed Baylor and the Confederates and would soon provide the Rebel forces with reinforcements.[14]

Quickly, the news that Baylor had entered and secured Mesilla reached Smith and the miners at Pinos Altos. One of the miners in the camp, Tom Helm, proposed that the Arizona Guards join Baylor's battalion. There was a heated discussion in the camp; some favored joining Baylor and the Confederacy, others argued for linking their fight with the Union side, and still more simply wanted to sit the war out. The men who favored the Union side left Pinos Altos and moved north to join Canby's troops. For the miners left at Pinos Altos, the threat of Indian attack was foremost in their minds. With the federal troops gone, the men were more concerned with protection from the Chiricahua than what side to take in the coming Civil War.[15]

The Pinos Altos miners decided to travel to Mesilla and talk with Baylor and the Confederates about frontier protection, the invasion of New Mexico, and where they fit into Rebel schemes. Fifty men left Pinos Altos, not for the express intention of joining the Confederate invasion, but simply to talk and determine what course of action they should take. Captain William Polk "Gotch" Hardeman of Baylor's company met the men at Picacho (about ten miles from Mesilla) and escorted the miners into the city. The Rebel occupiers of Mesilla were eager to receive any volunteers and heartily welcomed the Pinos Altos miners. Baylor's second in command, Major Edwin Waller, assembled the entire battalion at full attention on the main street of the town. The Confederates cheered Smith and his compatriots as they rode up the street. Waller had also arranged for Baylor and Frank DeRider, the superintendent of the Overland Mail station in Mesilla, to welcome the miners with rousing speeches. After the speeches, in Smith's words, the miners were "turned loose on the town." After leaving their horses and pack animals in the care of the Overland Mail station, Ignacio Barella, Antonio Anchito, and Juan Vega (whom Smith described as "the high cockelrums" of the town) invited the potential soldiers to a fandango, where "whiskey, mescal, and champagne were as free as water." Smith further remembers the night as one that was "a dandy . . . the Senoritas came out in their best harness . . . and every one of us fell in love."[16]

After consulting with Baylor, the Arizona Guards agreed to muster into the Confederate service, but they had to reorganize as a unit.

The ensuing election installed, according to Smith, Tom Helm as captain and Smith became the commissary sergeant. However, the official Confederate records of July 18, 1861, list Thomas J. Mastin as the captain of the Arizona Guards, a position that he presumably held when he was killed at the battle of Pinos Altos in October 1861. The Arizona Guards agreed to fight Union troops if needed, but their primary duties would be to keep the road to Tucson and the mines at Pinos Altos free from Indian attack, a duty that led to the Battle of Pinos Altos in September–October 1861. They were also assigned duties as general scouts, primarily since they knew the country better than any of Baylor's men, and their knowledge of Spanish and Apache was a great benefit. But before the Arizona Guards could return to Pinos Altos, they had to help repel a federal attempt to retake Mesilla.[17]

On July 25, Major Isaac Lynde and 380 US troops left Fort Fillmore with the intention of dislodging the Confederates from Mesilla. Lynde led his troops to the outskirts of the city and sent a message to Baylor demanding an immediate surrender. Baylor replied with a succinct statement full of the bravado that he was famous for: "If you wish the town and my forces, come and take them!" Lynde then formed a skirmish line and ordered a bombardment of the town. But the Confederates had ably fortified the city, while the desert sand and surrounding cornfields further impeded any advance Lynde had planned. After four or five rounds of artillery, Lynde sent a 250-man cavalry unit to within 250 yards of Mesilla and prepared to attack the Rebel fortifications.[18]

The show of federal force did not intimidate Baylor's Texans or the Arizona Guards. Firmly entrenched behind the strong adobe walls of the city, the Confederate force opened fire on the Union cavalry as soon as it came into range. The fire was very effective, with four Union soldiers killed immediately, and forced the Union cavalry charge to withdraw. Lynde tried to reform his command, but eventually decided to return to Fort Fillmore. Baylor sent a small force out, including Hank Smith and the Arizona Guards, to pursue cautiously the fleeing Union soldiers. Smith recounted that Baylor ordered the soldiers to "not push them too fast as we might run into an ambush in the bottoms of the Rio Grande." Presumably, Baylor simply wanted to make sure the federals did not reform and make another attack. The Guards returned to Mesilla on the evening of July 25 and prepared for another attack on the twenty-sixth, an attack that Lynde had no intention of making.[19]

However, the first battle of the invasion of New Mexico had cost the Confederate forces twenty horses, killed when a Union artillery barrage had struck a corral. Most of these animals had belonged to the Arizona Guards, and they received permission to replace their mounts and procure additional horses if possible. The Guards traveled south to the small village of Chamborino, near the Texas border. Smith was evidently placed in charge of this expedition and tried to buy horses, but the residents and merchants of Chamborino remembered Smith: "suspicioned [sic] our purpose and didn't care to talk horse, and they asked exorbitant prices for what they didn't want to sell." Frustrated by the turn of events, Smith and the guards fashioned another scheme.[20]

The guards returned to Mesilla, reported their efforts, and proposed their new idea, an idea that would allow them to find mounts at no cost and also hinder the enemy. The federal troops at Fort Fillmore had a good supply of horses, and the guards asked for permission to travel to the Union post under the cover of darkness to steal the mounts. At first Baylor rejected the plan but later agreed if it was carefully managed. The guards were to go forward with the plan only if they could do it without raising an alarm and without firing a shot. If that could not be accomplished, they were to return to Mesilla.[21]

The miners made contact with a pro-southern Union soldier at Fort Fillmore and, with his assistance, put their plan in motion. The guards divided themselves into two groups, one that moved quietly into the herd and another that watched the Union troops for movement. Fortunately, the Union troops guarding the horses were engaged in an intense card game and never heard the approach of the guards. Taking the federal guards by surprise, Smith and his men charged the Union camp and, without firing a shot, took the entire contingent prisoner. The raiders gathered both their prisoners and the horses and rode back to Mesilla. The guards had secured eighty-five cavalry horses and twenty-six mules. As a reward, Baylor allowed the raiders to pick their own mounts from the group. Smith chose the cavalry commander's horse, stating, "You bet he was a dandy."[22]

The loss of the horses was a huge blow to Lynde's troops, and it set up a blunder that would lead to a disastrous defeat. Lynde correctly assumed that Baylor would attack the next morning, and he ordered Fort Fillmore abandoned and the troops to proceed north to Fort Stanton. The federal soldiers set fire to the fort and marched out at dawn. When he arrived, Baylor ordered one contingent to put out the fires and the remainder to pursue the retreating federals. Smith stayed at the fort and helped fight the fires, but after subduing the blaze he and the rest of the Confederate forces moved out to catch the main company.[23]

The Union soldiers had not counted on the severity of the southern New Mexico desert. After only fifteen miles of marching, the hot and thirsty Union army disintegrated. As they overtook the stragglers, the Confederate forces encountered no resistance and began to take prisoners. Smith surmised why many of the Union soldiers had difficulty facing the desert. "We found some of the guns loaded with whiskey and a good portion of the soldiers drunk and begging for water." Another Confederate soldier, Private Milam Taylor, remembers seeing "stragglers of the enemy along the roadside almost dead from fatigue, thirst, and hunger." The retreating federal column also included 108 women and children, but many of the husbands and fathers, in fear and desperation, had abandoned their families and left them to the aid of the Confederate column. Another Confederate soldier, who had encountered dozens of these helpless victims along the route, called it a "pitiful sight."[24]

All that remained for the Confederates was to take the main Union column. As Baylor and his men approached the federal troops near San Augustin Springs, the Union soldiers tried to form skirmish lines with the aim of engaging Baylor's cavalry advance, but it was a half-hearted effort. Finally, with his options exhausted, Lynde and his one hundred remaining men surrendered. Baylor had accomplished what Sibley and his superiors felt would take many more men and much more time—he had secured southern New Mexico and with it a base of operations for the invasion of the northern reaches of the territory. The capture of the Fort Fillmore troops also allowed the Confederates to replenish their meager supplies and arms. The Rebels collected hundreds of Springfield rifles, ammunition, and more than $9,000 in treasury notes. Baylor and his troops returned to Mesilla and prepared to set up a political regime.[25]

Although John Baylor eventually declared himself martial governor of the Confederate Territory of Arizona with the capital at Mesilla, Hank Smith and the Arizona Guards would not spend much of their time fighting federal troops nor would they participate in the infamous Sibley invasion of northern New Mexico. While Baylor and the main Confederate contingent were concerned with organizing a new territory, the real threat of an Apache war remained. Mangas Coloradas and Cochise had not abandoned their war against white settlers, and the Mescalero peoples of southern New Mexico and western Texas were also taking advantage of the war to raid settlements, wagon trains, and supply lines. Smith and the Arizona Guards returned to the Pinos Altos region and to protect the territory's western frontier.[26]

Anticipating Henry Sibley's arrival in New Mexico, the Confederate command reorganized Baylor's troops. Baylor attached Smith and the Arizona Guards, along with three volunteer units from Arizona, a group of Arizona Rangers from the Mesilla Valley, and an excess from two other units, to his command. Baylor's troops had two duties—act as a rear guard for the invasion of northern New Mexico and protect the western frontier from Indian attack. There was not much need for a rear guard action since the Union troops had abandoned all the posts in southern New Mexico and western Texas, leaving the region to the Confederates. The Arizona Guards traveled back to Pinos Altos, where in late September and early October they fought the Battle of Pinos Altos, defeating a Chiricahua force.[27]

While the Arizona Guards were fighting the Chiricahua at Pinos Altos, the Confederate dreams of a southwestern empire took some serious blows. The earlier hope for control of California faded when reports arrived of five thousand California volunteers enlisting in the Union Army; other reports claimed US troops would soon invade Arizona in an ultimate attempt to dislodge Baylor from Mesilla. Also, the Union commander in New Mexico, Colonel Edward

S. Canby, had amassed 2,500 men at Forts Craig and Union and expected more reinforcements from Kansas. To make matters even worse, there had been no word from the expected Rebel volunteers from the mines of Colorado and Nevada. There was also a critical shortage of hard currency to purchase supplies since Mexican and New Mexican merchants were very reluctant to accept Confederate scrip. Pay for soldiers was also becoming increasingly infrequent. Baylor expected these problems to dissipate with Sibley's arrival with Texas volunteers, money, and supplies, but the situation demanded that they hurry.[28]

Baylor felt that his situation was growing tenuous. His spies reported on October 23 that indeed Canby and his forces would arrive in Mesilla by November 1 and would be augmented by the converging columns from California and another Union force from Forts Union and Yuma. If these reports proved true, the Union troops would outnumber Baylor's Confederate defenders almost ten to one. Baylor should have waited and substantiated the reports, but it seemed that his worst fears had been realized—federal troops would arrive before Sibley and the Texans. Baylor panicked and made an embarrassing blunder.[29]

The newly self-installed governor immediately issued orders to move his army and what few supplies remained out of danger. All food and fodder was removed from the storehouses at Fort Bliss to Forts Davis and Quitman in Texas. The citizens of Mesilla, who had shortly before hailed the Confederates as conquering heroes and promised to defend their cause to the death, reacted hysterically to the news, many fleeing to Mexico. Baylor's short conquest of southern New Mexico appeared to be in peril.[30]

Baylor, at first, seemed somewhat defiant, arguing that this was but a temporary setback, and with Sibley's arrival the Confederates would again occupy the Mesilla Valley and prepare for a larger invasion of the Southwest. But just two days after receiving the reports of Union movement, much of that bravado was gone. Baylor sent dispatches to his superiors in San Antonio pleading for reinforcements and predicting a disaster if they did not arrive within the next few weeks. He vented anger at Sibley for delaying his advance and even began disparaging the local Hispanic population whom he had earlier so carefully cultivated.

"The Mexican population . . . will avail themselves of the first opportunity to rob us or join the enemy. Nothing but a strong force will keep them quiet."[31]

But the Union offensive was a complete farce, and Baylor's reputation suffered greatly. In reality, the Californians were nowhere near Mesilla, and Canby's command was suffering from supply and morale problems every bit as severe as the Confederates. What had arrived at Fort Union was nothing more than a few supply wagons, but no troops or plans for an offensive. But Baylor and his troops had panicked on the basis of a few speculative reports and retreated to Fort Bliss to await the arrival of Sibley and reinforcements. Such a scenario represents the fragile nature of the Confederate position in southern New Mexico. Baylor's front-guard action had probably arrived in New Mexico too quickly, and only a series of fortuitous events had allowed them to gain control of the region with a force that was not truly sufficient to hold the area against a formidable Union force. The Rebels probably instinctively knew that their advance had been much too easy and were now seeing the specter of the tough fight they had anticipated.[32]

After only a few days at Fort Bliss it became apparent that there was no Union offensive on Mesilla; Baylor returned to Confederate Arizona. But the panic along the Rio Grande had caused him a great deal of embarrassment. The self-promoted hero of the Fort Fillmore fight and the self-appointed governor of Arizona now seemed to be a bumbling coward. The *Mesilla Times* excoriated the Confederate leader in an editorial. "Such a stampede never was witnessed, save at Manassas. Only we had a Manassas . . . without a fight or even sight of the enemy." Baylor, frustrated and looking for a scapegoat for his blunder, found one in Hank Smith's Pinos Altos home—butcher Anton Brewer.[33]

Brewer was the meat supplier for the Pinos Altos gold camps, meat that he brought from Mesilla and the Rio Grande Valley. Brewer had apparently stolen Confederate property left behind during the retreat to Fort Bliss, and Baylor quickly moved to take action against the butcher and deflect attention from his own situation. He ordered Hank Smith to Mesilla and gave him and fifteen of the Arizona Guards the task to pursue and arrest Brewer. Smith and his detachment made a half-hearted effort to bring in their friend. Brewer had a three-day head start as Smith and his men left Mesilla. When the group reached Picacho, they found a "big fandango in progress," and they decided to join in the festivities. In the revelry, they forgot all about their pursuit of Brewer. Reluctant to bring in their friend to begin with, Smith and his men returned to Mesilla the next day and reported that Brewer had escaped. An enraged Baylor stripped Smith of his sergeant's rank and locked the entire squad in the guardhouse.[34]

The Brewer episode began Hank Smith's disillusion with the Confederate cause. Probably never a "true believer" to begin with, Smith began to doubt the wisdom of fighting in the Confederate army. Baylor released Smith and his companions from the guardhouse after a few days, but he did not allow Smith to return to Pinos Altos. He appointed him as his orderly, either as a way to keep an eye on the young miner or perhaps as a method of punishment for the Brewer incident. Whatever the reason, the position gave Smith a firsthand view of a notorious incident involving Baylor and the editor of the *Mesilla Times*, Robert Kelly.[35]

It was Kelly who had written the acidic editorials in the Mesilla newspaper concerning Baylor's embarrassing flight from Mesilla in early November. Baylor had demanded an apology for the editorials, but Kelly had refused. On December 12, Baylor was leaving the adjutant office when he saw Kelly on the street. Baylor stepped back into the office, asked his new orderly Hank Smith to hand him a rifle, and moved to confront the news editor. Baylor again demanded that Kelly retract his statements, and when Kelly again refused, he swung the rifle at the newsman's head. Kelly avoided the blow, but Baylor tackled the man and pinned him to the ground. In fear for his life, Kelly reached for a knife in his belt, but before he could bring it up, Baylor drew his pistol and demanded that Kelly drop the knife. The next action is a matter of conjecture. Smith reported that Baylor again demanded a retraction, and with Kelly again refusing the Confederate commander shot him in the face. Other witnesses stated that the gun discharged during a struggle with Kelly over the knife.

Baylor immediately surrendered to his second in command, Major Edwin Waller, and later prepared for a murder trial because Kelly died two weeks after the incident.[36]

By most accounts, the trial was a sham. Smith was called to testify and reported the conversation between the two men and the circumstances of the shooting. Others testified as to the ongoing trouble between the two men and the "court decided that Colonel Baylor was justified in doing what he did." The residents of Mesilla demanded a civil trial, but Confederate authorities, citing martial law, denied the request. The shooting, combined with the heavy-handed tactics of Baylor and the Confederate occupying force, cooled the residents of Mesilla's ardor for secession. The treatment of Grant Oury, Mesilla's representative to the Confederate Congress, also caused Mesilla's residents to question the wisdom of Confederate control. The Richmond body had seated Oury but gave him no official status. Oury pressed the Congress for official recognition of the region as the Confederate territory of Arizona to no avail. He loudly protested the inaction and criticized Confederate treatment of the region's loyal residents. Baylor rewarded Oury's diligence with removal and, in a move that looked like cronyism, elevated his protégé and attorney general, Marcus MacWillie, to the position of territorial representative. Baylor was wearing out his welcome in Mesilla, and the once loyal city now had great doubts about the wisdom of joining the Confederacy.[37]

Smith got the opportunity to return to the Pinos Altos region shortly after Baylor's exoneration. With the bulk of the Confederate troops congregated around Mesilla, the Chiricahuas used the absence to continue their war against the white invaders. After the Battle of Pinos Altos, the Indians under Mangas Coloradas and Cochise attacked at least two freighter trains and then moved on to rest and wait for an extended campaign in the spring, with Cochise's band traveling to Mexico and Mangas Coloradas's group going to the Mogollon Mountains near the Gila River. An aging Mangas Coloradas may have contemplated peace with the white settlers in early spring of 1862. According to some accounts, he sent an emissary to Pinos Altos in March to request peace negotiations. But by that date, Colonel John Baylor, famed hater of Indians, was in the area and preparing to enact his signature brand of Indian policy—total extermination.[38]

In mid-February 1862, Baylor had decided to punish the Apache raiders. As Sibley's force moved out to face Canby's Union army, Baylor and a force that included Smith and the Arizona Guards went west to confront the Chiricahuas. Baylor was determined to force the Apaches into a decisive, crushing engagement that would eliminate the Chiricahua threat. An insight into Baylor's mindset is contained within his instructions to the Arizona Guards. After returning from a reconnoitering mission, Baylor issued his notorious order that was so insensitive and brutal that even his superiors in Richmond objected to its language (largely since Union propaganda would use the order to convince Native Americans to resist any Confederate overtures). In a perfect representation of Baylor's policy of genocide that he carried with him from Texas, in March 1862, he instructed Captain Thomas Helm of the Arizona Guards at Pinos Altos:

> The Congress of the Confederate States has passed a law declaring extermination to all hostile Indians. You will therefore use all means to persuade the Apaches or any tribe to come in for the purpose of making peace, and when you get them together kill all the grown Indians and take the children prisoners and sell them to defray the expense of killing the Indians. Buy whiskey and such other goods as may be necessary for the Indians. [I] look to you for success against these cursed pests who have already murdered over 100 men in this Territory.[39]

Baylor's force soon found itself in a frustrating and fruitless march through rugged mountains in search of an elusive prey. After arriving at Pinos Altos, Baylor sent his men to a nearby spring, where Hank Smith reported no sign of the Chiricahua band. Baylor returned to Pinos Altos and began a march south into the Big Burro Mountains where Smith again reported no indications of Indians but "plenty of turkeys and black tail deer." Replenishing their food supplies with the wild game, the column continued on to Steins Peak and the Peloncillo Mountains. Again, finding no sign of Indian camps Baylor decided to continue his pursuit into Mexico.[40]

Baylor moved his troops in a hard march into northern Chihuahua. Near Corralitos, the men found fresh evidence of the Chiricahuas, but according to Mangas Coloradas's biographer, Edwin Sweeney, this was more than likely members of Cochise's band instead of Mangas Coloradas's (who was probably still in the Mogollon Mountains). As the men made camp, Baylor sent his scouts, including Hank Smith whose knowledge of Chiricahua and Spanish was invaluable, into the surrounding villages. The villagers testified that there were Apaches in the Tres Hermanos Mountains, fifty miles north and just inside the borders of Confederate Arizona. But pursuit was not an option; the Confederate mounts were exhausted, and a fast sprint to the mountains was just not possible. Instead, Baylor decided that the best option was to move around the Indians through a series of easy marches, even though such a strategy lengthened the journey by over a hundred miles. After four days of travel, Baylor's troops were in a position to attack. Dividing his forces, the plan was to attack the Indians from the rear first, drive them into the canyon, and force an ambush. At dawn the Confederates started the attack.[41]

But the Chiricahuas had again fooled the Rebel forces. Sometime during the march, they had escaped south back into Chihuahua. Baylor began another quick pursuit into Mexico, pushing the meaning of "hot pursuit" to its physical and tactful limits. He followed the Indians to Corralitos, a remote mining community west of Casas Grandes. He overtook the Apaches and they sought refuge in the hacienda of José María Zuloaga and sent out peace overtures. Baylor, true to his reputation, refused any talk of surrender and ordered his men to attack. The resulting gunfire killed most of the Indians and scattered the rest. Baylor's men captured a few stragglers, and the Confederate commander ordered them immediately executed. Satisfied with his victory and feeling that his reputation was restored, Baylor returned to Mesilla, and Smith and the Arizona Guards made their way back to the mines of Pinos Altos.[42]

Baylor's raid into Mexico produced diplomatic aftershocks that rocked the region and the Confederacy's reputation. The gore and mayhem was shocking even to the notorious anti-Indian sensibilities of Baylor's contemporaries and jeopardized Confederate diplomatic efforts with Mexico. R. L. Robertson, a US Mexican consul, was amazed at Baylor's audacity and predicted that the Mexican authorities would be furious. Mexicans, given their history, were already suspicious of any Texan incursions into their country and aggressively protested Baylor's invasion. For them, this went far beyond any definition of "hot pursuit" and constituted an outright invasion of their sovereign soil. Baylor, and most of the Confederate leadership, seemed oblivious to the charges.[43]

Baylor's actions against the Chiricahuas would ultimately prove his downfall. In the end, it was the order he drafted to Captain Helm and the Arizona Guards that ended his Confederate career. In the order, Baylor had stated that the CSA Congress had "passed a law declaring extermination to all hostile Indians." In truth, the Confederate government had passed no such law and was committed to the old US policy of pacification. The scheming Baylor had blatantly misrepresented the government's policy. Baylor was also disillusioned after Sibley had arrived in New Mexico and left him in little more than an administrative position in Mesilla. Bitter at being placed in the rear, Baylor left a letter of resignation on his desk in Mesilla, boarded a stagecoach, and headed off for Texas. Eventually, citing his order to Helm and other actions, Baylor was relieved of all authority and mustered out of the Confederate Army. Like Sibley later, Baylor's dreams of a southwestern Confederate empire died due to blunders and miscalculations in New Mexico.[44]

Smith and the Arizona Guards were now back in Pinos Altos, but their situation was anything but desirable. Supplies were low, the mines were all but abandoned, and the threat of Indian attack was imminent. Smith does not record, nor do any records reflect, his movements throughout the spring of 1862. Knowing the territory from his work on the Overland Mail route, Smith very well could have accompanied a detachment of the Arizona Guards that occupied Tucson in March of that year. But he just as well could have spent the spring in Pinos Altos as part of the small contingent protecting the mining camp from Indian attack.[45]

The entire Confederate dream of a southwestern empire was also falling apart in the spring of 1862. In April, Sibley's invasion of northern New Mexico ended when he lost his supply train at the Battle of Glorieta Pass and made the decision to quit New Mexico. By the end of June, Sibley was back in Texas and only a small

Confederate force of just over four hundred men in Mesilla, along with the Arizona Guards in Pinos Altos and Tucson, were all that remained of the Confederate invasion force. After Sibley's departure, Colonel William Steele ordered the Arizona Guards to abandon Pinos Altos and Tucson and move back to Mesilla to augment his force in the southern New Mexican city. The Pinos Altos men, who had originally come to the region to make fortunes in the mines, now found themselves as a part of a failed vision of empire and part of a beleaguered rear guard confronted with horrible conditions.[46]

Smith and the Arizona Guards, most of whom had reluctantly joined the Confederate adventure, now had to confront another obstacle. Their small force faced considerable odds. To the north were the Union forces that had pushed Sibley out of the region, and fast approaching from the west were the federal troops from California, sent specifically to expel the Confederates from Mesilla. Steele moved his forces, now including Hank Smith, to Fort Fillmore, which he felt could be better defended. The Confederates had only thirty days of food, were low on ammunition, and had little transportation. These meager supplies were saved for the regular troops, and the independent companies, such as the Arizona Guards, were forced to forage and steal from villagers.[47]

Smith, and no doubt many of his companions, was thoroughly disgusted with the situation. Smith had always been an unenthusiastic Rebel. He was no southerner; indeed, he had never lived in the South outside of the small period he spent in Missouri. He probably understood few of the issues of the war and had joined the Confederate Army simply out of loyalty to his mining compatriots. Added to such conditions was the fact that Smith had not been paid for his service to the Confederacy and had suffered the indignity of a demotion in rank. With the federal troops from California nearing Fort Fillmore, Hank Smith decided that he no longer wished to be a Confederate soldier. On July 4, 1862, with only a musket and a small ration of food, he deserted. His decision was not an isolated incident. Records show that when the remnants of the Arizona Guards reached San Antonio after the abandonment of New Mexico, eleven men had deserted. Ironically, one of Smith's reasons for leaving was probably to keep from being forced to retreat to Texas, a state where he would eventually spend most of the remainder of his life.[48]

No doubt aware that Colonel Steele and the Confederates would soon have to abandon Fort Fillmore, Smith hid in the scrub brush surrounding Mesilla for five days—until Steele ordered a withdrawal. But Smith did not act like a man on the run. He moved into Mesilla and tried to blend in to the population. His ruse did not work; when the Union army entered the area shortly after the Confederate withdrawal, he was arrested for disloyalty to the United States. Smith could have faced serious consequences, but the federal authorities needed men who knew the territory. Given a choice of imprisonment or taking the oath of allegiance to the Union, Smith never hesitated and pledged service to the United States. Major James Carleton, the Union commander in Mesilla, hired Smith as a contract scout, a civilian employee of the Union army. However, Smith spent little time as a scout; his main service was in an occupation that he knew well, a freighter and forage contractor.[49]

Shortly after his arrest and release, Smith reported to Lieutenant Sidney DeLong, the quartermaster at Mesilla. DeLong gave Smith the express contract for the Rio Grande. He directed Smith to deliver supplies, mail, and forage between Mesilla and the Rio Mimbres, pick up another load and continue on to Fort Cummings before returning to Mesilla. Smith was well suited to the job—he was well acquainted with the territory and had experience as a freighter and forager. For the remainder of the war, he carried the mail, dispatches, and supplies between the forts and camps of southern New Mexico. During this period he also met a man who would become a lifelong friend, confidant, and business partner, Charley Hawse. While Smith never recorded how he met Hawse, records from other sources indicate that Charley Hawse, a native of Maine, was with the Union army in New Mexico. Hawse was apparently a sergeant of the guard and accompanied Smith during his express duties. Hawse and Smith lost track of each other after the war but were reunited in Fort Griffin in the 1870s. The work as an expressman for the Union army gave Hank Smith valuable contacts that he would use to his advantage when the war ended.[50]

Chapter 4

There is no record of when Smith ended his service to the Union army, but it was probably after July 1865 when the last of the Confederate armies surrendered. Rather than continue his career as a miner, with the added experience as an expressman after he left the Confederate army, he decided to attempt to make his fortune as a freighter. It was a business that he had first learned as a wagon master on the Santa Fe Trail and seemed the best business opportunity at the time. Although he would begin his business in the Mesilla Valley, Smith eventually made the move to Texas. He had avoided this move when he deserted the Confederate army, but he then began a new business in the growing city of El Paso.

Hank Smith relaxes on the porch of the Cross B Ranch. Photo courtesy of Southwest Collections/Special Collections Library, Texas Tech University, Lubbock, Texas

Hank Smith as "Old Shatterhand." Old Shatterhand was a fictional character in western novels by German Karl May. Photo courtesy of Lubbock Avalanche Journal

Smith built the Rock House when he took possession of the Tasker Ranch. In addition to being the Smith residence, it was also used as a post office and a general store. Photo courtesy of the Crosby County Pioneer Memorial Museum

Pinos Altos, NM, photo from *The Pinos Altos Story,* by Dorothy Watson.

Hank Smith, pioneer cowman of Texas, sitting on the brow of a hill in Blanco Canyon overlooking his ranch. Cross B Ranch, Texas. 1909, Gelatin dry plate negative, 5 x 7 in. Erwin E. Smith Collection of the Library of Congress on Deposit at the Amon Carter Museum of American Art, Fort Worth, Texas, LC.S6.490, © Erwin E. Smith Foundation

Hank and Elizabeth Smith sit for a formal photograph. The photograph was taken in Fort Griffin where the couple lived before establishing the Cross B. Photo courtesy of the Crosby County Pioneer Memorial Museum

5

Hank Smith, Texas Entrepreneur

The Civil War, like it did for so many people, represented an "interruption" in Hank Smith's life. Before the war began Hank Smith had left the Gila River region with the idea of moving to the El Paso area and a career as a trail trader or expressman. He got sidetracked when he followed his penchant for grasping at business schemes and potential riches as he answered the siren call of gold mining in Pinos Altos. The period he spent in Pinos Altos led to what in hindsight may seem adventurous, but it was actually a time of conflict and tension for Smith. He became caught up in a war that he did not truly support and watched his vision of gold riches disappear under the clouds of war and conflict with Native Americans. So, one could not blame him for wanting to begin with a new direction, a direction that resurrected his earlier dreams of a freighting and trading business.

Hank Smith had no desire to return to the mines of Pinos Altos. Like so many independent miners in the West, he had never prospered in the gold mines. With the war over, he turned back to his earlier vocation, operating a trade train over the Santa Fe Trail. Although the records of this period in Smith's life are haphazard at best, he apparently left his southern New Mexico base in the summer of 1865 and traveled to Kansas City, the starting point of the Santa Fe Trail. Selling himself as an experienced trader, Smith gained employment as a wagon master for a mule train with a full consignment of goods; he made two trips a year between Kansas City and Tucson. To reach Tucson, he traveled the original trail into Santa Fe, where the train sold and traded what goods it could, loaded more goods in the New Mexico capital, and continued on to Mesilla. From

Mesilla, the train turned west for the three-hundred-mile trip to Tucson, where Smith sold the remainder of the goods and purchased more goods for the return trip to Kansas City.[1]

In 1865, the Santa Fe trade was beginning its slow decline. The primary culprit for the decline was the railroad. But in the first few years after the Civil War, with the extending of the railroads west, the Santa Fe trade experienced a boom. As the railroads pushed west establishing new railheads, these new towns often became locales for new warehouses and commission establishments. As the lines stretched west from Kansas City, the route to Santa Fe also became shorter, cutting out fifty to seventy-five miles of often slow travel. The business of freighting was also changing. Instead of trains pulled by oxen, those drawn by mule or horse trains were the norm. Also, shorter, quicker trips meant merchants had a longer season to receive goods, and they no longer had to buy six months' or a year's supply of goods at one time. It also meant that the initial investment was smaller, which also meant more competition. More and more men entered the business of hauling freight from rail terminals to places not served by railroads.[2]

Smith left no financial records of his period on the Santa Fe Trail in the 1860s, so his profitability on the route cannot be known, but he did leave some handwritten notes containing a few vignettes about his time on the trail. Traversing the Santa Fe Trail in the 1860s meant mundane activities most of the time combined with short bursts of adversity. The most common hardship by far was Mother Nature. Smith had to contend with flooded rivers, the death and loss of mules and horses, insect infestations, poisonous snakes in the night camp, and the passing of buffalo herds. During one trip he waited for five days on the banks of the Neosho River (in southeastern Kansas not far from the Oklahoma border) for it to recede enough for crossing. On a spring trip, a swarm of lightning bugs stampeded the stock animals and he had to spend two days bringing the teams back. In 1868, on the Pawnee Fork of the Arkansas River, he came upon the southern buffalo herd. He had to stop his train for four days to allow the herd to pass, but he did kill five of the animals and stored enough meat to last the remainder of the trip. Snakes slithering into the camp at night were a constant problem. During one spring night, Smith claims that he awoke with four "of the vile creatures" on his bedroll. When he reached the trading post at Fort Bent, he immediately purchased a cot that raised his bed at least two feet off the ground.[3]

Smith continued his trailing business on the Santa Fe Trail until 1869. Although he did not leave any business records, the sheer numbers of traders on the Trail in the years immediately following the Civil War must have made for fierce competition. Such a situation meant the traders had to cut expenses greatly on their trips, which made profits scarce. As a sole operator, Smith no doubt was at a competitive disadvantage. Also, the life of a trader on the Santa Fe Trail, two round trips a year in excess of two thousand miles each way, must have been brutal. Faced with declining profits, Smith decided to try his business luck in another location, a place that he had set his sights on a decade earlier, the then small hamlet of El Paso, Texas.[4]

The original settlement at El Paso del Norte was on the Mexican side of the Rio Grande. But in the decade after the Mexican-American War, the United States began an extensive exploration of West Texas, primarily in the interest of building a transcontinental railroad across the region. One of the most promising routes was the so-called southern route, which would extend across central Texas to El Paso del Norte and on to the Pacific coast. Interest in establishing such a route resulted in two expeditions in 1849, which resulted in the opening of two roads from San Antonio to El Paso. With an overland route established, the inauguration of mail service from San Antonio through El Paso and on to Santa Fe began. The increased volume of mail brought the need for a post office in 1852 for El Paso. First named Franklin to avoid confusion with the Mexican city of El Paso del Norte, a small settlement grew around the post office, and the American city of El Paso saw its beginnings.[5]

In 1858, John Butterfield's Overland Mail Company began operations and exerted the greatest impact on the development of El Paso until the arrival of the railroad. Butterfield contracted Anson Mills to build a station at El Paso, which opened in September 1858, only three days before the first stage arrived. The El Paso station was the largest and best equipped in the Butterfield line and was the most imposing structure in the small town. Mills next took on the task of platting and laying out the city of El Paso, which he completed early in 1859. By 1861, El Paso had four hundred residents, people who saw the city as a gateway to mines in New Mexico, or who were lured there by trade with Mexico by 1861. It had become nationally known as an important way station on the southern route to California.[6]

The trauma of the Civil War and the immediate years afterward blunted El Paso's growth and prosperity. The war years allowed the farmers and ranchers of the Mesilla Valley in New Mexico the opportunity to capture trade from El Paso. Due to the city's relative isolation, prices were high and supplies sporadic. But in the late 1860s and early 1870s, some people were optimistic enough about the border city to open businesses. By 1872, El Paso had at least three saloons, three hotels, three doctors, six lawyers, and a Masonic Lodge building. It also had newly recommissioned Fort Bliss. The post was a good source of income for traders and merchants.[7]

The disastrous end of Henry Sibley's abortive attempt to take New Mexico for the Confederacy left Fort Bliss in poor condition. The retreating Confederate army had stripped Fort Bliss of anything of value and had left behind a barren, deserted post in ruins. In October 1865, the US government moved in to rebuild the vital frontier outpost and poured well over a thousand dollars into the local economy with the employment of painters, masons, and laborers. The expenditures continued throughout the remainder of 1865 and into 1866. In April 1866, the newly recommissioned fort welcomed the arrival of three companies of the US Fifth Infantry. But Fort Bliss, located on the banks of the Rio Grande, was prone to flooding. To alleviate the situation and locate Fort Bliss to a more secure location, the federal government in 1867 leased one hundred acres on the high terrain east of El Paso for a new post. The building of the fort began another flurry of economic activity and population growth in El Paso.[8]

Also, in the years after the Civil War, Anglos began to dominate El Paso's business, political, and social life. They were a strong-willed and individualistic group who sought wealth and power. Most were engaged in merchandising, freighting, mining, or the law. They brought with them the mechanisms and practices of American government, which allowed them to manipulate elections and control local officials. They hired new arrivals to work in their businesses, intermarried with the Hispanic elite, and made valuable contacts with the *Paseño* aristocrats across the Rio Grande. El Paso was not a farming frontier like those that so dominated other regions of the West. It attracted wage earners and potential businessmen, not the landless poor in search of property. At its core, El Paso was an exploitative region in which self-seeking men sought to acquire wealth through merchandising, freighting, or mining.[9]

With a few wagons and teams secured from his Santa Fe Trail business, Smith arrived in El Paso sometime in the spring of 1869. El Paso's location made it an ideal headquarters for freighters, with Mexico just across the Rio Grande, an army post nearby, and trailheads north to New Mexico and west to California. Smith had made valuable contacts during his tenure as an expressman for the federal forces in Mesilla, and upon his arrival in El Paso immediately made his way to Fort Bliss to inquire about army hauling contracts. The fort had already awarded the contracts for 1869, although they welcomed his bid for 1870. But he learned that a contractor was needed to furnish hay and wood for Fort Quitman, a small outpost seventy miles southeast of El Paso on the Rio Grande. Smith went to Fort Quitman, made a bid for the job, and in August learned that he had acquired a new business. The army agreed to pay him seven dollars a cord for wood.[10]

Smith's contract was not set to begin until January 1870, but the isolated fort needed hay and wood immediately, and Hank agreed to begin operations. In late 1869 he received word that he also had received the contract to provide hay and wood for Fort Bliss, greatly expanding his operation, which meant that he had to hire employees to help with the business as well as purchase more wagons and mule teams. Smith's business records indicate that over the course of the next two years he hired over fifty employees, bought at least twenty wagons, and at its peak used more than one hundred mules and wagons. His new business proved to be very capital intensive. Despite the appearance of a thriving concern, Smith was barely breaking even. To help balance the books, he took on more contracts, anything in an attempt to accumulate more capital. It would be a practice that he would continue throughout his life. While supplying the two forts was his biggest enterprise, he also took on private contracts to supply wood and hay to other El Paso area businesses. In 1869–70, he filled an order with the El Paso Mail Company for forty-one tons of hay at $10.50 a ton ($1.50 more than he was charging the US Army) and supplied P. Mooney with 181 cords of wood at $3.25 a cord. But despite taking on as much business as possible, Smith could still not make up the capital expenses of starting his hauling business; he found it necessary to enter into a limited partnership with Luis Cardis and Frank R. Diffenderfer.[11]

Cardis was a well-known and politically connected citizen of El Paso. An Italian immigrant, Cardis's fluent Spanish and Old World sense of style brought him a large following among El Paso's Hispanic population. Anglo politicians regularly enlisted Cardis to help attract support among the Mexican American population. During this era of Reconstruction, Cardis, an ally of Democratic politicians, used an unusual style of political partisanship. He was a close friend and ally of Father Antonio Borrajo, the parish priest of San Elizario. Cardis had Borrajo threaten to excommunicate any Catholic who voted the Republican ticket. Cardis entered the Texas legislature in 1874 and continued as a political power broker in El Paso until he was killed in 1877 during the El Paso Salt Wars.[12]

Diffenderfer was the brother of David Diffenderfer, the former United States consul for El Paso del Norte. The brothers operated a general mercantile business, and Frank was also the post sutler at Fort Bliss. Frank was the person to whom Smith delivered the wood and hay at the fort. Both Cardis and Diffenderfer provided the capital for Smith to invest to expand his business, primarily for supplies and to meet a payroll. The agreement that Smith signed with the two men allowed him to purchase supplies from the Diffenderfer store and draw money from Diffenderfer accounts to pay his employees as long as there was sufficient hay and wood in the stockpiles to cover the expenses. This agreement eventually caused Smith to overdraw his accounts, and both Cardis and Diffenderfer brought legal action.[13]

Smith's hauling and forage business grew and offered promise for profits for the first half of 1870. But there were indications of problems, difficulties that eventually caused him to lose the contracts with the forts and caused the business to fail. Although hay was abundant in the bottomlands along the Rio Grande, anyone with knowledge of the El Paso area knows that finding wood in the deserts of West Texas is a difficult proposition. Most of the wood in the area was mesquite, a hot and fast-burning wood but not a reliable lumber source. Part of Smith's agreement with Diffenderfer allowed him to take over a previous contract for the taking of wood from land owned by Pablo Alvaranda in Mexico. The contract called for only one hundred cords, but Smith exceeded that amount by the middle of 1870. Alvaranda sent Smith a strongly worded letter instructing the wood entrepreneur to cease any more operations on his land and demanding payment for the overage. Smith had other sources; records indicate that he hauled wood from the Quitman Mountains southeast of El Paso and that he also made ventures into the Sacramento Mountains of New Mexico, near the small village of Dowlin's Mill (known later as Ruidoso). But Smith had constant difficulty finding enough wood to fill his contracts, and the expense of sending crews to distant locations proved overwhelming.[14]

Hank Smith had made the mistake that many businessmen make when they enter a lucrative but underserved market—he expanded too much, too quickly. By the end of 1870, he could no longer meet his obligations. Cardis and Diffenderfer suspended Smith's credit line, and soon officials at Fort Bliss began pressuring him to pay his expenses to the sutler's operation at the fort. At the end of 1870 he lost the contracts for Forts Bliss and Quitman and had to try to settle his accounts solely from the smaller contracts he had in the city. Diffenderfer and Cadis dissolved their partnership with Smith in January 1871 and refused to accept any more orders from him. Cardis pressed Smith for payment for the supplies and cash he had advanced to Smith's employees. Smith was in no position to settle his debts, and in May 1871, Cardis sought a judgment for debt in the amount of $1,120. The court ordered Smith's business property seized for payment, and Cardis received eighteen mules, three wagons, and four saddles, along with various other tools. Diffenderfer demanded Smith return all wagons, harnesses, saddles, and tools. Smith's business was crumbling, and he was on the verge of insolvency.[15]

Hank Smith's only solution was to find someone with capital to join his fledgling business and rescue him from bankruptcy. He found just such a person in Roe Watkins, an El Paso merchant who had dealings with Frank Diffenderfer and no doubt knew of Smith's situation. Although it cannot be substantiated through any records, one can assume that Diffenderfer was the person who put Watkins in contact with Smith. In order to alleviate his debt, in August 1871, Smith agreed to take on Watkins as a partner; in return the El Paso merchant agreed to pay Smith's debt to Diffenderfer. After Watkins acquired Smith's business, he had no trouble providing a promissory note to the Fort Bliss sutler for the wagons Smith had surrendered to settle his debt. Watkins provided Diffenderfer with a $1,992 note payable in six months. Diffenderfer released the wagons and teams and a new partnership of Roe Watkins and Hank Smith entered the freight business.[16]

Although he was free from his debt to Diffenderfer, Smith's business was still in peril. He owed accounts at Fort Quitman, and without the fort contracts the pair had to find business elsewhere. Watkins and Smith journeyed to Fort Quitman on August 26, secured a load of corn from the firm of G. W. Wahl, and agreed to ship it to W. E. Friedlander and Company in Fort Stockton. Smith must have been convinced that Watkins could handle the load since he accompanied his new partner only as far as Fort Davis and left Watkins alone to continue to Fort Stockton. Smith then went back to Fort Quitman to see if he could settle his debts. While still at the isolated outpost, Smith received a letter from Watkins reporting that he had dropped the corn in Fort Stockton and made arrangements to continue on to San Antonio with another shipment. Watkins hired twenty recently discharged soldiers, paying them fifteen dollars each in advance. Three weeks later, on October 1, Watkins arrived in San Antonio. In his correspondence to Smith he reported that the army had offered a contract to haul lumber to Fort Concho at fifty dollars per one thousand feet, but that he did not believe the price would yield a profit and felt it would be better to travel to Indianola, on the Texas gulf coast, and try to find freight headed for Forts Stockton or Davis or "some other place in the up country." He also revealed that he thought it prudent to travel to Mexico during the winter to buy cheap corn.[17]

Smith must have contacted Watkins and told him to remain in San Antonio and that he would soon join him. He must have settled his debt problems at Fort Quitman by the end of 1871 since he arrived in San Antonio in late January 1872. The Watkins-Smith train was in financial difficulty. Obviously, the pair never made it to Indianola; they remained in San Antonio in April 1872. One can only speculate what the two partners discussed during this period. It was becoming increasingly obvious that the business could not sustain two men with an adequate income. A business recession gripped the nation during 1872 and Hank Smith was still saddled with debt and legal problems from his earlier sole proprietorship. Watkins had also been away from El Paso for three months and needed some resolution to the partnership. In the end, Smith probably decided to accept the only solution he had available—turn the entire business over to Watkins. In a contract dated April 19, 1872, the two men agreed to dissolve their partnership. Smith turned all assets of the partnership over to Roe Watkins and gave Watkins a power of attorney to act as his agent in settling all debts and liens against him in El Paso. He also agreed to pay Watkins for any expenses or settlements incurred. Watkins must have successfully paid most of the debt with profits from the freighting business; the only debt that remained a year later was the $1,992 note to Frank Diffenderfer. On April 3, 1873, Diffenderfer traveled to Fort Griffin, Smith's new home at the time, and received from Smith a check for the full amount, plus 4 percent interest.[18]

Hank Smith's first Texas business venture had ended in failure, but he quickly moved on to another enterprise. Smith's true talent lay as a wagon master, not in running a large business. He must have been quite the salesman since he never lacked for securing enterprise. His only trouble had been in keeping it operating. But the competition on the San Antonio to El Paso road was fierce, so Hank Smith began looking for other opportunities to perhaps less serviced areas. It did not take long for him to find another wagon train to lead. Within a week of dissolving his partnership with Watkins, Smith accepted a job leading a load of freight consigned to Fort Griffin, Texas. This would prove to be the beginning of a new venture for Hank Smith, one in which he would find his first taste of success and that would eventually lead him to operating a ranch on the South Plains of Texas.[19]

Fort Griffin was one in a line of forts that guarded the supposed "frontier" line of settlement against Indian raids. It lay near the Clear Fork of the Brazos River in Shackelford County. In the 1860s, the region was a sparsely populated land where a hardy vanguard of entrepreneurs primarily operated cattle operations. But that began to change in the late 1860s when the US Army decided to locate a new post, soon to be named Fort Griffin, in the area. Quickly, the fort attracted a congregation of civilians. A civilian population was vital to any army post since the residents filled many vital jobs. The presence of the army was also attractive for the number of contracts they could award. Fort Griffin gave out contracts for providing the fort with wood, beef, and milk. The most coveted job was as the post sutler. Of course, army forts provided a ready market. Lonely, isolated young men "needed" outlets that saloons and bordellos provided. However, the stereotype of debauchery is more often than not, in the least, exaggerated. Nonetheless, by 1870 the community around Fort Griffin, referred to by the soldiers as the Flat, boasted three saloons.[20]

Hank Smith's migration to Fort Griffin can be seen as part of a larger migration within Texas, although Smith approached the area from a different and more circuitous route. From the mid- to the late 1870s, white Texans began to push out of the wooded environs of east and central Texas onto the prairie. In the decade of the 1870s, soldiers, settlers, and buffalo hunters finally defeated Native Americans, either through outright extermination or through forcing them onto reservations in Oklahoma. The defeat of the Indians (along with the almost total extermination of the buffalo) meant that white settlers could now move into the previously isolated and uninhabited area, setting the stage for explosive growth in West Texas.[21]

Common stereotypes of these "frontier" towns, including Fort Griffin, characterize them as wild, wide-open communities full of rowdy and murderous cowboys, gamblers, and outlaws. In many historical interpretations, Fort Griffin fits just that image. While drunken cowboys and frontier violence is certainly a portion of the history of the region, it is by no means even the most pervasive or transforming part of West Texas' past. Such a theme is one of the central theses of Ty Cashion's work *A Texas Frontier: The Clear Fork Country and Fort Griffin, 1840–1887*. Cashion responds to those who believe in, and even revere, the stereotype that while "the popular perception ... in this area is most typically recalled in stories of Indian raids, cavalry campaigns, wild 'nights on the town,' vigilante hangings, buffalo hunts, and cattle drives, such colorful chapters have overshadowed the mundane activities that also formed the regional experience." Cashion's study demonstrates that the so-called mundane activities were some of the most important facets of the history of the region. Of the activities, the formation of viable businesses had perhaps the greatest effect on the settling of the region. Hank Smith became one of the first businessmen in Fort Griffin and one of the most successful.[22]

While the fort made Fort Griffin possible, buffalo hunting made it a population center and prepared it for growth. The great buffalo hunts in northwest Texas only lasted five years, from 1874 through 1879, but they were a boom industry for the region. By 1875, throngs of buffalo hunting outfits had descended into the Clear Fork region, and with them also came freighters, gamblers, saloons, and prostitutes. One longtime resident recalled that by the spring of 1875 "rows of hides, tier upon tier" were stacked on the edge of town. Wagonloads after wagonloads full of hides made their way into the town, sometimes as many as one hundred a day.[23]

Businesses that specialized in the buffalo trade were the ones that were the most innovative and well capitalized. A successful buffalo hunter had to possess considerable skill as both a marksman and businessman. The business of buffalo hunting was much more tedious and commercial than its romantic image. The owner of a buffalo outfit had to organize and win the respect of a crew of independent frontiersmen and negotiating for the sale of hides required considerable skill and tenacity. But while the buffalo hunters were in effect the labor force of the hunting boom years, it was the men who fueled the boom, the outfitters and the merchants who dominated the economic aspects of the hunt. When the boom in Fort Griffin began, merchant Frank Conrad shipped in as much powder, guns, and ammunition as he could find. In January 1877, the *Fort Worth Democrat* reported that he posted a $4,000 day, $2,500 of it in guns and ammunition. Conrad also supplied the hide wagons and built a virtual "machine" in the city with his profits.[24]

For the men who became buffalo hunters, the business provided a source of income in an economy that often held little promise. A Texas farmhand in those years could expect to earn no more than twenty dollars a month for a life of hard labor. Cowboys and ranch hands were scarcely better off. But one "robe quality" buffalo hide could bring $3; even less desirable hides of smaller cows earned as much as seventy-five cents. Men with no experience could earn top wages as skinners, drivers, camp tenders, or any number of jobs. With such potential, hunters and outfitters flocked to the region to join the slaughter. Hides stacked over ten feet and covering more than ten acres surrounded Fort Griffin.[25]

Hank Smith came to Fort Griffin in its early formative time and eventually shared in the profits of the buffalo hunt years. But when he arrived in the area, he found only a fort and a vast, sparsely populated rolling prairie. Although there were a few scattered dwellings around the fort, most of the residents in the region were stock raisers, using the vast open grasslands to feed their herds. The tradition of instability on the rolling plains—Indian warfare, Civil War interruptions, and the harshness of the land—had left the area without a viable commercial center and very few services. It seemed an unlikely place to begin a new business, but Hank Smith saw potential in the Clear Fork country.[26]

Smith made a remarkably quick trip from San Antonio to Fort Griffin. If the dates written down in his records are correct, it took him only seven days to make what was normally a ten- to twelve-day journey. He unloaded his freight at Forth Griffin's sutler store and learned that the hay contract for the fort was open. W. H. Hicks had supplied the fort with hay but no longer wanted the job; Smith quickly secured the contract. He returned to San Antonio to finalize his business with Roe Watkins and also arranged to lead another load of freight for Fort Griffin. He hired eight local men to accompany him to the fort and cut the hay for the contract. Smith had also started to be more precise in his business dealings. He had each of his new employees sign an employment contract that explicitly detailed each man's wages and how he would pay them. The men signed on for five months of work as teamsters and herders. Teamsters would earn twenty dollars a month and herders fifteen. Smith must have made a decent profit on his first Fort Griffin load as he agreed to pay each man their first month's wages in advance. After that, the contract stipulated that each man would receive two dollars a month while in Fort Griffin, with the remainder of their pay held in security until the end of the five-month period. If any man left before the end of the five months he forfeited all pay due to him. Smith had problems securing labor in El Paso, and for this venture, obviously, he wanted to ensure that each man remained until the contract was fulfilled.[27]

Smith arrived back in Fort Griffin in the beginning of June 1872 and immediately began to cut hay and fulfill the contract. Grass was plentiful on the plains near the fort and along the bottomlands of the Clear Fork, and Smith reported that it was a much easier endeavor than what he had faced in the desert surrounding El Paso and Fort Quitman. However, he encountered difficulty in July when an Indian raid into his camp caused him to lose twenty-six of his mules. But Smith was determined to make this venture a success and immediately found another supply of mules; at the end of the contract in October he made his full delivery.[28]

With the onset of winter, Smith cut back his hay-cutting business. His San Antonio employees must have gone home; their names never appear again in any of Smith's business records. He had the contract for 1873 secured, but he had bigger plans: a more stable business that could grow with the town. But he lacked capital, and with his income stagnated for the season he needed steady pay that could generate cash flow. So, he turned the day-to-day operations of his hay and hauling business over to a newly arrived Scottish immigrant, James Boyle, and took a job as a clerk at Conrad and Rath's General Merchandise Store. For Smith, the job represented a way to save money so he could eventually begin a new business or expand his current one. He also started to take jobs as a carpenter and continued to help his new partner hauling hay and wood for the fort and the residents of Fort Griffin. But his association with James Boyle would prove to have other advantages as well. James Boyle's sister, Elizabeth, would soon join her brother on the Texas prairie.[29]

Smith and Boyle continued to fill the hay contract for the fort through 1873. Smith also continued clerking at Conrad and Rath's as well as building corrals, stock pens, and doing other projects for the residents of the area. The skill he acquired as a machinist's apprentice in Germany also helped him earn extra income, as he became a sort of "handy-man" for the soldiers at the fort and for other people around Fort Griffin. Through these various enterprises, Smith was finally accumulating a cash reserve that he could use to begin another operation. The end of 1873 began another chapter in Hank Smith's life: James Boyle's sister, Elizabeth, arrived in Fort Griffin.[30]

When he arrived in Fort Griffin, Hank Smith was thirty-five years old. In a relatively short period of his life he had already seen more of the world than the vast majority of people did during their lifetime. It had been a busy life for the German émigré, a life that he had spent trying to make a living and simultaneously accumulate some degree of wealth. His life had probably not worked out as he had originally planned, but he was, perhaps, now in a position to see some success. But his had also been a life of upheaval, one that had him moving constantly and never settling in one place. While this type of existence may have suited a young man, it did have its drawbacks; Smith had never found the time to marry, raise a family, and settle in one place. Perhaps he saw the opportunity to correct that missing piece of his life in Fort Griffin; maybe he saw the rustic frontier village as a place where he could operate a successful business, begin a family, and finally find stability. Perhaps he was simply a romantic man and was looking for the right woman.

Elizabeth Boyle was born in Dairy, Ayrshire, Scotland, in 1848, into a landed family; her father had inherited the Fleshwood estate. She was an educated woman for her era, attending school in Scotland for seven years at the Presbyterian school Blairmains. But opportunities in Scotland during the 1860s and 1870s were limited, and all four of her brothers had left Scotland for Texas. She was encouraged to make the journey as well, primarily to "keep house for them." The Boyles first settled in Missouri, were soon joined by their father, but moved to Texas after hearing of "better opportunities." Elizabeth took a train to Dallas in September 1873 and boarded a wagon to Fort Worth, where James met her and brought her to Fort Griffin.[31]

The plains of Texas must have been quite a shock for the young Scottish woman. The wagon trip from Fort Worth to Fort Griffin was the first time Elizabeth had "camped out by wagon." She arrived in Fort Griffin just in time to attend the New Year's Ball, the social event of the year in the town. At the ball, James introduced her to his business partner, Hank Smith. Smith was no doubt heartened to meet a woman of Elizabeth's age (twenty-six) who was unmarried. A very quick courtship began, one in which "at first Hank came to see me every Sunday; then he came twice a week." Boyle's Hank Smith was ready to take a wife and settle in Fort Griffin. On May 19, 1874, Hank and Elizabeth married. The ceremony must have been quite an event for the young town. Fort Griffin's commander, G. P. Buel, and many of the officers attended, as did most of the ranchers and merchants of the area.[32]

Smith continued his job at Conrad and Rath's after the marriage and his hay business with his new brother-in-law. James tended to much of the cutting and loading, and Smith, who helped when he could, took care of securing additional business. The two partners also began supplying wood to the new businesses in the town surrounding the fort. But soon Smith and his new wife began discussing another business venture. Fort Griffin was growing and was becoming a regular rest stop for settlers, hunters, and cowboys headed west. In 1874, D. M. Dowell of Kentucky learned that the army had built Fort Griffin on sections of the Peters Colony, land that he had purchased in 1858. Dowell announced that he intended to survey and divide his land, sell lots, and establish a community near the fort and obtain the county seat of newly organized Shackelford County.[33]

The residents of Shackelford County were set to vote on a county seat in November 1874. The surrounding cattlemen had put forward an unpopulated spot eighteen miles south of the army post as the county seat, but Dowell's new "town" presented a formidable alternative. It was closer to the fort, it contained residents, and it had the backing of Dowell and his capital. Hank

Smith did not mention this controversy in any of his records or notes; his business was with the fort and what few residents lived in the "Flat," not on Dowell's land. However, he and Elizabeth had discussed opening a hotel near the fort, and a good location for such an enterprise would be in Dowell's town. More than likely, Smith favored Dowell's land for the location of the county seat.[34]

The army did not favor Dowell's proposal. Army officers saw a new town near the fort as a potentially volatile situation for the men at the fort; they began to try to sabotage Dowell's efforts. Their campaign was aided when post trader W. B. Hicks produced a lease that Dowell had supposedly signed giving him control of the land surrounding the fort. When he heard of the lease, Colonel Buel threatened to remove all the settlers from the land and prevent the erection of any new buildings until Dowell showed he had full legal authority over the land.[35]

The controversy over legal title to the land around the fort swayed the election. With the question of who controlled the land in question, some settlers decided simply to "squat" on Dowell's land until Fort Griffin soldiers removed them at gunpoint. The incident probably swayed some undecided voters and, in the end, the cattlemen's location won a close election and became the town of Albany. But by 1875, Dowell and Colonel Buel had worked out a lease agreement and the town of Fort Griffin would become the commercial center of the area.[36]

With the establishment of Fort Griffin and the legal wrangling completed, Hank and Elizabeth Smith proceeded with their plans for a hotel. Smith bought land at the base of Government Hill directly on the road leading to the fort. On one lot he constructed what would be a temporary home. On another he began building what would become the Occidental Hotel. The couple also decided to begin a family, suggesting that Smith was beginning to think of Fort Griffin as a stable home. Their first son was born on May 22, 1875, but only lived until October of that year. The Smith's also suffered another family loss that year when James Boyle drowned in the Clear Fork when his hay wagon turned over near the river. They moved into the Occidental in July and began operations.[37]

The Smith's hotel was a success from the start. Since Fort Griffin had become an important crossroads, the Occidental had a high occupancy rate. Fort Griffin became a supply center and stopping point for buffalo hunters, people moving west, and cowboys seeking jobs. But the Smiths were not content with operating just a hotel. Under the strict guidance of Elizabeth, the Hotel Occidental included a dining room stocked with the freshest food available, many of the vegetables grown in Elizabeth's garden, and a steady supply of fresh beef that Smith secured through Conrad and Rath's general store. On Sunday, Elizabeth set the tables with fine china and silver—a chance to offer a bit of refinement in a frontier existence. Hank also operated a bar in the hotel, but it was an establishment that he refused to call a saloon and was supposedly reserved for the convenience of hotel patrons. Regardless of his purposes, Smith's records reveal that the bar was one of the most profitable operations at the Occidental. He also operated a wagon yard and livery stable at the rear of the hotel, another business that was a profitable addition to the new town. Hands from the neighboring ranches boarded their horses in the stable when they came to Fort Griffin for relaxation and revelry. Smith rented, sold, and repaired wagons in the yard. According to the *Frontier Echo*, in February 1877, "the livery stable is patronized better than any layout in town."[38]

With the death of James Boyle, Smith gave up his hay contract and concentrated on operating the hotel and a freighting business. Hank operated the bar and the livery stable and also continued to do carpentry. Elizabeth ran the hotel and the dining room and also took over much of the business's bookkeeping, working even during a pregnancy when the Smiths welcomed a son, George, in October 1876. The Smiths became prominent citizens within the growing town. Although hotelkeepers and saloon owners in some frontier towns did not find social acceptance, the Smiths had no trouble in that area and even acquired the affectionate monikers of "Uncle and Aunt Hank Smith." The Smiths prospered, and for the first time Hank Smith began to carve out a comfortable existence. It was the first business that he had operated successfully since he had left Ohio. Perhaps Hank had learned the nuances of business, and those lessons made him a success. But another theory cannot be discounted; this was the first business he had operated since marrying Elizabeth Boyle. She took over much of the day-to-day finances for the couple, and the educated Scottish woman proved an adept accountant.[39]

When the bison slaughter overtook Fort Griffin, Hank Smith was well positioned to profit. As a freighter, he found more business than he could reasonably afford to take hauling hides to railheads for shipping. Smith did not have the size or capital to become one of the largest haulers, but he found a niche. Men with such means, such as Frank Conrad, could ship 200,000 pounds of hides in one run. Smith's records indicate that he never operated or led more than ten wagons at a time, but with the trade in hides so large and lucrative, he could count on his wagons always being full and his profits high.[40]

Although the Smiths were prospering in Fort Griffin, Hank Smith continued to seek business opportunities. He gained a reputation as a man who would listen and consider any business scheme and, if he found it within reason, accept a proposal. In December 1875, a man with just such a proposal came to Fort Griffin and made a pitch. The result would change the Smiths' lives and ultimately bring them to an isolated and virtually forgotten section of Texas.

Charles Tasker was the scion of a Pennsylvania steel family who had dreams of operating a "grand ranch" in the West. Don Biggers was twelve when Tasker arrived in Fort Griffin, and the Philadelphia native made quite an impression on the young man. Biggers was a frequent guest of the Smiths, and he was at the Occidental when Tasker made his entrance. "A new hack, drawn by a pair of fine-blooded horses, drew up before the hotel. Its occupants were a liveried driver and a well-dressed man and woman. Following the hack were several wagons loaded with provisions and supplies, particularly good wines, whiskies, and cigars. Charles Tasker owned the small caravan." Tasker began making inquiries, spent a great deal of money, and let it be known that he was looking for a buffalo ranch. Tasker also had a penchant for gambling, and the card sharps in Fort Griffin recognized a mark every time he entered a saloon; Tasker left a good deal of his money at the tables.[41]

Tasker had never operated a ranch and had no idea how to begin. What he had was money and a dream. A stock operation in the nineteenth century was no easy task. Ranchers faced drought, disease, and the vagaries of the market. Cattle ranching had ruined a good number of men wiser than Charles Tasker. But Tasker would not be deterred. After asking residents about who would be the best man to scout the areas west and locate a ranch, Hank Smith became the obvious candidate. Tasker asked Smith to locate a proper location for his "grand ranch," a spot with "available water, good grass, and without the distraction of a town." Smith agreed and recruited Judge John Schimerhorn and three other men to find Tasker a ranch.[42]

Smith never revealed why he chose Blanco Canyon as the location for Tasker's ranch. Perhaps it was the first place he came to that fit all of Tasker' s qualifications. Another theory is that Tasker wanted to operate a buffalo ranch and, according to Don Biggers, Blanco Canyon was the perfect place for just such an operation. "This was on the headwaters of Blanco River (Creek) . . . on the south the canyon comes to a narrow width. Some three miles north of this spot was another narrow place. By building stone walls across the north and south ends of the canyon . . . the buffalo park would be complete." But the buffalo hunters had practically killed the entire southern herd, and there were not enough buffalo left to stock a ranch.[43]

Smith returned to Fort Griffin and reported his find to Charles Tasker. Tasker and a companion started to Blanco Canyon to see the site. Apparently Tasker approved of Smith's selection. The land in Blanco Canyon was part of the Eastland County school lands. Tasker acquired four sections (2,650 acres) and made plans to begin his ranch. He hired hands, raised a cattle herd, and sent them to start work on the new ranch. Tasker intended to live on his new estate and needed a home, so he contacted Hank Smith about whom he could hire to build what he would call "Hacienda de Glorieta." Smith must have seen the Philadelphian as an easy mark, just as the Fort Griffin gamblers had earlier. He introduced Tasker to men in the town he could hire and ordered and took shipment on lumber and tools from Fort Worth to build the Taskers' dream estate. Smith, who of course knew the freighting business and had access to wagons and teams, agreed to transport the goods to Blanco Canyon, all for a price to be paid after workers completed the house and Tasker's uncle forward ed funding. Ultimately, Hank Smith made over $11,000 in outlays to Tasker, money that represented more than his entire assets in Fort Griffin.[44]

Smith was not the only resident of Fort Griffin or Texas who had extended Tasker credit. Tasker had gambling, liquor, clothing, and other debts in many areas of the state. He had previously run up enormous debts in Dallas and also had other unpaid bills in Fort Worth. While the work was proceeding in Blanco Canyon, he continued spending money in Fort Griffin, living, in Smith's words, "the high gate route." All his creditors believed that Tasker's funds were unlimited, backed by the uncle in Philadelphia. Merchants, saloonkeepers, and gamblers just extended him more credit, limits that he soon exceeded again and again.[45]

While he may have exaggerated his wealth, it was obvious from the home he was building in Blanco Canyon that Tasker had expensive taste. In the summer of 1876, he sent teamsters, cooks, masons, hayers, hunters, and herders all to Blanco Canyon to transform a wilderness area into a glorious estate. He hired experienced Irish stonemasons to quarry and fashion the exterior of the home. The walls of the house were twenty-two inches thick, and the outside dimensions were forty feet by nineteen feet. The house reached a height of eighteen feet. It had two rooms downstairs, three rooms upstairs, and four fireplaces. It had intricate masonry work and expensive wood interiors. But the Rock House (as it became known after Hank Smith acquired it) was not enough for Charles Tasker. While work was still proceeding on the original dwelling, Tasker had workers lay the foundation for a larger and grander residence, the culmination of Hacienda Glorieta. The foundation for this building was still in place when Smith moved his family to Blanco Canyon in 1878, and if Tasker had ever finished it Smith said it would be a "veritable mansion in the western wilds."[46]

Tasker had hired a Fort Griffin man, Charlie Smith, as his ranch foreman. He had a reputation in Fort Griffin as a bully and a thief and had endured several altercations with post soldiers and law enforcement authorities. The new foreman hired a few hands and began the trip to the ranch. As Tasker began to prepare to move to Blanco Canyon, it became apparent that his funds were not unlimited. When the workers who had built his home in the canyon returned to Fort Griffin for their pay, Tasker pleaded a lack of funds, asking the men to wait until he could secure money from his uncle in Philadelphia. But his uncle had learned of Tasker's debts in Dallas and Fort Worth, and while it is not clear if he knew of his plans in Blanco Canyon, he decided to stop his nephew's "allowance." Tasker was also under investigation for fraud in Dallas. The merchants and suppliers in Fort Griffin suspended Tasker's credit privileges and demanded payment. Hank Smith was caught in a dilemma; he had laid out $11,000 to Charles Tasker, money that he had to recoup or his business in Fort Griffin would suffer, even fail. The only hope he had of ever getting his money back was if Charles Tasker could operate and profit on the Blanco Canyon ranch. Faced with little choice, he loaned Tasker $500 more so the Philadelphian and his wife could make the trip to the South Plains, although the Rock House was not fully completed.[47]

Blanco Canyon would ultimately prove to be a good location for a ranch, but Charles Tasker was not the right man to make it work. Smith, who probably had more contact with Tasker than anybody in Fort Griffin, had a low opinion of the man who now virtually controlled his future. At different times he called him a "rattle-brain," a "spendthrift," and "a gambler and a sorry character generally." Tasker was not interested in operating the ranch; he was much more interested in traveling back to Dallas or Fort Worth, and when he was at the ranch he concentrated on plans for his grand mansion. His hands had brought six hundred head of cattle to Blanco Canyon, and the men and Charlie Smith completely operated the ranch. But soon Charlie Smith and Tasker's employees faced the same fate that Tasker's builders had—Tasker never paid them. Charlie Smith, known as "Chief Red Mud" in Fort Griffin, was a brutish man who began to demand his wages. Tasker rebuffed Chief Red Mud, and the bully began to threaten and terrorize Tasker, demanding his pay. Frightened, broke, and seemingly out of options, Charles Tasker decided to abandon his dreams of a "grand cattle ranch" and leave Blanco Canyon. Tasker convinced a young ranch hand, John Birdwell, to guide them off the ranch in the spring of 1877, and he left his Glorieta Hacienda, never to return. Tasker eventually abandoned his wife in Dallas, and, faced with mounting debts and legal proceedings, went to Mexico. He was eventually captured and spent two years in prison.[48]

With Tasker gone, his creditors moved in to repossess any asset they could find. Hank Smith was in a long line of creditors, but his situation was perhaps the most desperate. He had bargained his entire future on Charles Tasker and had poured most of his assets into furnishing the materials for Tasker's home. Smith tried to contact Tasker and his uncle in Philadelphia to recover his money but could never find Tasker, and his uncle refused to pay. With no other choice, Hank Smith filed suit against Tasker in Shackelford County and received a judgment for $11,000. But other creditors had picked over most of Tasker's assets except for the one thing nobody seemed to want—Tasker's Blanco Canyon ranch and home. Reluctantly, Hank Smith took possession of Tasker's home, four sections of land, and six hundred head of cattle in payment for debt. Hank Smith, who had been a teamster on the Santa Fe Trail, a sideshow attraction for a California tavern, a gold miner, an expressman, and a soldier, was about to become a cattleman.[49]

Hank Smith had no choice. Every penny he had made in Fort Griffin was now tied up in another man's failed dream. The Smiths closed the Occidental Hotel, loaded their children and their possessions into a wagon, and on a brisk November day in 1878, set off down the Mackenzie Trail for Blanco Canyon. Smith did not record his or his wife's emotions as they left the only home they had shared together. Elizabeth had buried a son, brother, and her father in Fort Griffin. She had enjoyed operating the Occidental Hotel and the dining room. For her, Blanco Canyon must have seemed the edge of the world and a place far from her life in Fort Griffin, to say nothing of her Scottish home. It was in Fort Griffin that Hank Smith had finally found business success and a place that he had obviously hoped to live the rest of his life. This was not anything like the previous times in his life when he had moved on for another opportunity; he now had a family to take care of, two young children and a wife. It was probably a brooding and subdued Hank Smith who hitched his team and moved west. Hank Smith would eventually become perhaps the most celebrated and revered man in his new home, but when he set out for his new life it was with sadness and a foreboding sense of what lay in store for him in Blanco Canyon. He was starting over again.[50]

6

The Cross B Ranch

Crosby County's most prominent citizen, Hank Smith, had built a successful business in Fort Griffin; it was also there that he married, began a family, and hoped to find stability in his life. He was forty-two years old in 1878, and during a period in American history when most people never moved much farther than the immediate vicinity of where they had been born, he had traversed through much of the West, looking for a place where he could find economic security and a permanent home. But that life was behind him, and he and his wife were now faced with trying to make a life on a ranch in Blanco Canyon in an area that was isolated and virtually uninhabited.

Blanco Canyon is part of the Caprock Escarpment, an exposed edge of Cretaceous foundation rock that the early Spanish explorers referred to as the Llano Estacado. The White River carved the canyon; it runs southeast for a total of thirty miles to its mouth. Its average depth is about fifty feet, but as it nears its mouth, eight miles southeast of present-day Crosbyton, its cliff heights average three to five hundred feet and it is over six miles wide. The soil is predominantly loam, and scrub brush and grasses dominate in the area. When the Smiths arrived at the canyon, the river provided a dependable water source, there was adequate wood for fuel, and the walls of the canyon offered some protection from the winds, hail, and the sudden snowstorms that often threatened the region.[1]

While it was unpopulated when the Smiths arrived, the area was not unknown. The Coronado Expedition passed through and camped in the canyon in the 1540s. It had been a favored campsite for numerous Native American peoples for centuries, and buffalo hunters and other western travelers sought refuge within the canyon walls. It was also the site of the climax of Colonel Ranald S. Mackenzie's initial campaign against the Comanche in West Texas. In September 1871, Mackenzie received permission from General William T. Sherman to mount an expedition against the Kotsoteka and Quahadi Comanche bands, which had refused to submit to demands to return to the reservation in Oklahoma in the aftermath of the Warren Wagon Train Raid. Mackenzie gathered eight companies of the Fourth United States Cavalry, two companies of the Eleventh Infantry, and a group of twenty Tonkawa scouts at Camp Cooper in late September. The column set out on October 3, hoping to find the Quahadi village, including the Quahadi warrior Quanah. Quanah and his men were camped in Blanco Canyon, near the headwaters of the Fresh Water Fork of the Brazos, southeast of what would become Hank Smith's home. Mackenzie reached Blanco Canyon after four days of quick marching and established a base camp at the junction of the Salt Fork of the Brazos and Duck Creek. The next day the infantry were left behind at the camp while the scouts and cavalry continued on.[2]

Late on October 10, Quanah and a Comanche force stampeded through the cavalry camp, driving off sixty-six horses. The following morning a detachment of troopers set off down the canyon chasing a small group of Comanches who were driving several horses. Topping a hill in a rugged section of the canyon, the soldiers came upon a much larger Indian force waiting in ambush. In the ensuing skirmish, one trooper was killed and a detachment of five men held off the Comanches until the remainder of the troops escaped.[3]

Only the timely arrival of Mackenzie and the main body saved the men from total annihilation and forced the Quahadis to withdraw. The Comanches retreated up the bluffs, sniping at the troopers before disappearing over the Caprock. Mackenzie pursued the Indians for the next two days, forcing them to abandon lodges and possessions as they fled. He finally caught up to the Indians on October 12, but an unseasonable "blue norther" with snow and sleet prevented him from making an attack. After the storm passed, he continued the pursuit for over forty miles, before he halted the campaign and turned back. As it returned south, the detachment fought another skirmish in Blanco Canyon with a Comanche scout party. Wounded in this skirmish, Mackenzie eventually terminated the expedition and returned to Fort Richardson.[4]

Mackenzie regarded the campaign as less than successful. He and his troops had marched over five hundred miles and had accomplished not much more than to frighten one Comanche band. But he had done something that affected the Indians' psyche; a white force had penetrated the Llano Estacado, an area that had always been the Comanche's safe haven. The campaign served as a learning experience for Mackenzie and his troops, knowledge that allowed them to continue penetrating the Llano Estacado and deny it as a safe haven for Comanche bands. Three years later, almost exactly to the day, Mackenzie defeated the Quahadis in Palo Duro Canyon and virtually ended Comanche existence in West Texas and opened the area to white settlers, including Hank Smith.[5]

Smith had come to Blanco Canyon without his family in early 1878. Elizabeth was pregnant at the time with their second child Leila, who was born on September 12 of that year. But Smith did not come alone. At some point in 1877, Smith's longtime friend from his Civil War expressman days, Charley Hawse, had arrived in Fort Griffin. After the war, Hawse had found work with one of the buffalo hunting outfits, probably as a wagon driver for the hides and a skinner. Hawse had lost contact with Smith and had no idea that he was now living in Fort Griffin. Smith was glad to see his old friend, someone from his former life with whom he could share a bond and conversation. He convinced Hawse to quit the buffalo hunters and go to work for him at the Occidental Hotel and the livery stable. Hawse would show remarkable loyalty to the Smiths for the remainder of his life, never marrying and working side-by-side with the couple. Smith, Hawse, and John Henry Jacobs, who had previously been a buffalo hunter, left in March 1878 and set out to Hacienda Glorieta to find out exactly what Tasker had left and make preparations for the rest of the Smith family to arrive.[6]

Hank Smith had invested over $11,000 in Charles Tasker's dream and now had to find a way to recoup his money. He had never operated a ranch and probably had a very rudimentary knowledge about what a stock operation entailed. But he had been a store clerk and knew that the Llano Estacado was still an area for buffalo hunters, men who would need supplies and essentials and would welcome an outpost in the lonely country. So his first order of business for what he would call the Rock House was to open a general store, primarily for the buffalo hunters in the area. He had hired Jacobs and his wife to manage the store, and the men had loaded four wagons with implements to stock the new enterprise. Smith's ledgers reveal that he planned to stock the store with flour, dried beef, molasses, blankets, corn meal, dried beans, and salt, all items that the buffalo hunters would need. He would trade his goods for hides and buffalo tongues, goods that he could profitably send to eastern towns. During his initial years in Blanco Canyon the general store was a prime source of income for the family; it allowed Smith to slowly begin building his ranch operations.[7]

When the men arrived at the Tasker ranch, it was in a state of disrepair. Tasker had never had the Rock House completely finished, and the corrals were virtually unusable. It is likely that Charlie Smith and his ranch hands had stripped the operation of anything useful since Tasker had never paid them. Charles Tasker, of course, had been consumed with building his mansion, probably to the neglect of the operation. John Jacobs quickly got to work setting up the general store, while Hawse and Smith finished the interior of the Rock House and rebuilt the corrals and pens on the ranch. Smith stayed at Blanco Canyon until the late summer of 1878 and then left Hawse and Jacobs (whose wife had now joined him) at the ranch and traveled back to Fort Griffin to bring his family to their new home.[8]

The Smith family left for Blanco Canyon in November 1878. Hank and Elizabeth, along with their two-year-old son George and two-month-old daughter Leila, and a young African American employee who Elizabeth had hired to help with the children after a difficult pregnancy, made their way down the Mackenzie Trail to their new home. When they arrived, Charley Hawse and the Jacobses met them and helped everybody get settled. Jacobs had moved the store from the Rock House to a half-dugout that had served as a bunkhouse under Tasker. Elizabeth Smith remembers her first impression of her new home as positive. She described a wild plum thicket and large vines of wild grapes in a fertile valley full of grass and wild game. There is no doubt that she had a great deal of trepidation about moving her young children into such an isolated country, but the sight of such a lush valley surely eased her anxiety.[9]

But the Smiths and their employees were all alone. Their nearest neighbor was on Duck Creek, fifty miles southeast, and the nearest ranch was Jim Reid's property one hundred miles southeast in Stonewall County. There were a few buffalo camps spread throughout the area, the closest at Buffalo Springs in Yellowhouse Canyon. In Dickens County, to the east, were the Dockums, who would later operate a store and a post office. There were still some bison in the region, which brought hunters to Blanco Canyon and the Caprock. Smith also hunted bison in his first years in the canyon, but he soon realized that he could make more money on trade with buffalo hunters and the byproducts of the plains animals. He traded for hides and tongues at his store, which he could sell for good profits, and he also collected and shipped bones, which were left on the plains by the tons. He hauled wagonloads of bones to Fort Griffin and later to Amarillo and Colorado City after the railroad reached those cities. The bones, used as an ingredient in fertilizer, also provided Smith with cash income.[10]

Smith saw potential in his new home and almost from the moment he arrived in the canyon began promoting the area as prime for settlement. He and Elizabeth embraced the canyon as their home and prepared to make a life. He wrote and posted a letter to the *Frontier Echo*, which appeared in the paper's December 6, 1878, edition, extolling the virtues of Blanco Canyon and the South Plains. In the letter, he made sure that the readers of Fort Griffin's newspaper knew that Blanco Canyon was "finely watered by numerous lakes of pure fresh water" and had a source of water power at Silver Falls. He urged the federal government to discontinue the El Paso to Santa Fe mail route and instead run the mail through the Llano Estacado. Smith, like other promoters of the West, tended to exaggerate the area. He called the country "full of buffalo" and an "excellent place to raise hogs." He claimed that he had "killed yearling shoats which weighed nearly three hundred pounds each" a few days before and "they have not had one kernel of grain." This is probably a false claim. While Smith may have brought the hogs to the ranch in 1877, there is no mention of hogs in his inventory of 1877. If he brought the hogs with him in 1878, they would have just completed an arduous journey from Fort Griffin and, while it may have been a lush valley, Blanco Canyon was not nearly that fertile. Smith's true intentions were revealed in the last paragraph of his missive when he urged the *Frontier Echo*'s readers to come see the canyon and its wonders:

> I know you are fond of hunting and fishing and you will receive a hearty welcome And can indulge in the sports to your heart's content, as game is abundant, and as is fish, we can show you more of them, larger ones, and varieties innumerable than can be found in any country of earth. So come bring your fishing and hunting outfit—we will furnish all necessities including "red liquor"—enjoy a few weeks' sport and relaxation from business.

Even though he was alone on the plains, Smith was thinking about profits—he could be considered the area's first chamber of commerce.[11]

The Blanco Canyon region would prove to be ideal for raising stock. The presence of Comanche bands had kept white ranchers from the region largely until the 1870s, but once the US Army conquered, vanquished, and removed the Comanche to reservations, ranchers found the region and began a new era. For centuries Blanco and Yellow House Canyons were the prime winter buffalo pastures on the Southern Plains, full of canyon rimrock for shelter and abundant grass. Those same conditions also made it ideal for cattle. Elliot Roosevelt, Theodore Roosevelt's younger brother, described Blanco Canyon during the winter of 1876–77 as a place whose running stream attracted unbelievable swarms of deer, antelope, turkey, and bison. When Charles Tasker had sent Hank Smith to scout a location for a "grand cattle ranch," he had found a fine spot.[12]

Although the Smiths were determined to stay and prosper at their Blanco Canyon ranch, their first year was not without difficulties. Smith was learning how to operate a ranch for the first time, but, although there is no indication in the record, it seems that Charley Hawse at least had some experience with a cattle operation. While in Fort Griffin, Hawse spent most of his time working in the buffalo camps, but he very well may have also found employment on one of the many Shackelford County cattle operations. Hawse took the lead in tending the herd that first year, and Smith learned from him. The two new ranchers had the advantage of the canyon, as cattle did not tend to wander far and the pair could work the ranch alone. The Blanco Canyon store was also showing modest profits. Smith used the occasion of being called to serve on the Shackelford County grand jury to travel back to Albany in March 1879. He took his wife and children with him as well as three wagonloads of buffalo hides, bones, and other trade goods from the store. His ledgers indicate that he received $326 for his goods, a fair return for an isolated trade outpost.[13]

While the Smiths were away in Albany, the new ranch experienced the only true violent encounter during its existence. Perhaps aware that the Smith family was away and anticipating an empty house, a group of men robbed the Rock House and the store. The thieves apparently caught Hawse and Jacobs off guard; besides stealing some of the Smith's household goods, they also took Jacobs's boots and Hawse's watch and pistol. Elizabeth, who related the story, indicated that the thieves left the area and went to New Mexico, where they were soon killed. Mrs. Smith never named the criminals, but Hubert Curry has speculated that Charlie Smith and the remnants of Tasker's ranch hands were the perpetrators. He and some of the Taskers former employees lived a few miles away from the Smiths at the mouth of Blanco Canyon. Perhaps Charlie Smith still resented the fact that Tasker had never paid him and decided that he would take his revenge on the current occupants of the ranch. Also, the fact that he was never seen in the area again seemed to support Elizabeth's story.[14]

The subjugation of the Comanches meant that the western plains of Texas were open to settlement; a few settlers began to join the Smiths in the Blanco Canyon area. W. C. Dockum, a buffalo hunter, settled in Dickens County near the present-day city of Spur. Dockum and his wife soon opened a store. Much to Elizabeth's delight, a female neighbor, with whom she often visited, was nearby. Dockum eventually moved to the Quaker settlement at Estacado and served as a Crosby County judge.[15]

The first cattle operation (other than Smith's spread) in the

Blanco Canyon area was the beginning of one of the most celebrated operations in West Texas, the Matador Land and Cattle Company. When he took over Tasker's claim in 1877, besides the land Tasker owned, Smith received the right to lease Eastland County School land in the Canyon for eight years. He only had six hundred head of cattle and did not need the additional land so, in 1878, he subleased his holdings to two New Mexico cattlemen. After staying in the canyon for two years, the two men moved on to lands in Motley County and Dickens County and eventually sold the herd and land to H. H. Campbell. Campbell had taken a herd of longhorns to Chicago, and following the sale he formed a partnership with banker A. M. Britton, then returned to Texas to acquire a range and establish a ranch. After the initial purchase along the Pease River in Motley County, he began buying more land and also assuming the right to most of the free-range land in the region. Within a year, the partners brought in other investors, including Spottswood Lomax of Fort Worth. Early in 1879 the Britton-Campbell partnership was organized as the Matador Cattle Company and reported assets of $50,000. Smith and Campbell became close friends, and it was from Campbell that Smith learned many of the nuances of cattle raising.[16]

Campbell continued to operate the ranch for the partnership into the early 1880s. With a good water supply, the abundance of plains grasses, and the high beef prices of the period, the ranch prospered. Although the ranch was showing profits, Britton felt that the time was right to sell to a larger firm. During the 1870s and 1880s, venture capitalists in Great Britain became interested in American cattle operations and, in Dundee, Scotland, Britton found a group of businessmen eager to invest funds in American land and cattle. Britton sold the operation to several Dundee merchants, and the Scots incorporated the Matador Land and Cattle Company, a joint stock company that had control over more than 1,500,000 acres of rangeland and 40,000 head of cattle. Campbell remained as the ranch superintendent under the new owners and began an upgrade of the property that included fencing, windmills, and additional land. Campbell left the Matador operation in 1890, and Murdo Mackenzie, who further expanded and diversified the huge cattle raising operation, replaced him.[17]

Smith and Campbell continued to be good neighbors and often corresponded and socialized through most of the 1880s. But relations between the two and Smith's giant neighbor ranch became strained with the organization of Crosby County. Smith became the first tax assessor of the county and made an attempt to get a correct assessment of the number of cattle and improvements on the Matador for tax purposes. Matador officials objected to Smith's numbers and felt that he had greatly overcharged in the assessment. Matador officials, at the urging of the company secretary, Alexander Mackay, conducted an organized and furious campaign against Smith in the next election that led to his defeat. Although Smith claimed that he knew that Campbell did not have a large role in the effort, the two men's relationship faltered, and Smith, from that point, considered the Matador Land and Cattle Company political and personal rivals.[18]

After Matador's sale to the Scottish investors, Campbell's partners, A. M. Britton and Spottswood Lomax, began another cattle operation in the region, one that was not quite as successful. Britton and Lomax purchased 242,000 acres in what would become Dickens, Kent, Garza, and Crosby Counties and incorporated as the Espuela Land and Cattle Company. Eventually, the Spur Ranch (as it became known) covered almost 570,000 acres, including twenty sections of other public lands. Most of the cattle purchased to stock the range came from small ranchers who, no longer having access to an open range, had no choice but to sell out.[19]

Two years later, knowing from his experience with the Matador that English and Scottish investors had a great deal of interest in Texas cattle operations, Britton traveled to London and roused the interest of a group of financiers. The British interests purchased the land and livestock from Britton's company, and for the next twenty-two years the British syndicate operated Espuela, at times much their chagrin. Droughts and a less than favorable cattle market, as well as poor management, made the enterprise less than successful. Finally, in 1906 the syndicate sold Espuela for five dollars an acre, a price that included all cattle, horses, buildings, and equipment.[20]

In 1880 the Smiths acquired permanent neighbors within three miles of their ranch, the H-L Ranch under M. V. Blacker. In 1886

Blacker sold out to Tom Montgomery, who established the TM operation. Eventually, Montgomery bought or leased about one hundred sections in southeast Floyd County and northeast Crosby County and ran over four thousand head of cattle. Montgomery also became Smith's friend and took part in the organization of Crosby County.[21]

The largest operation in the immediate Blanco Canyon area was one that Smith never regarded as particularly good: the Two-Buckle Ranch operated by the Kentucky Cattle Raising Company. The ranch headquarters for the huge, twenty-thousand-acre outfit was only ten miles down Blanco Canyon from Smith, near Silver Falls. Although Smith occasionally did repair work and other odd jobs for the Two-Buckle, Smith and his neighbors did not share cordial relations. The Two-Buckle bought up much of the Eastland County School land that Smith had under lease, as well as other parts that Smith had used as free range. Smith always felt that the school district should have given him first shot at the land; in his mind, the Kentucky Cattle Raising operation kept him from expanding and restricted the use of a number of acres of valuable land. Perhaps Smith thought that the free-range era would simply last forever.[22]

Members of a wealthy Louisville, Kentucky, family, the Tilford-Johnsons, organized the Kentucky Cattle Raising Company in September 1882. Claude Tilford, a cousin of the controlling family, became the resident manager of the ranch. The headquarters of the Two-Buckle Ranch was located just above Silver Falls, and at its height it contained more than 13,000 head of cattle. But the Two-Buckle became a victim of an overproduced cattle market and the horrible 1888 winter. Exorbitant costs for windmills, freight, and farm supplies, combined with cattle losses during the winter and large principal and interest payments, doomed the operation. By 1899 a Louisville bank foreclosed the property and the ranch was broken up and sold.[23]

Another stock operation that had unsatisfactory dealings with the Kentucky Cattle Raising Company ultimately became one of the most successful ranches in the region. Captain James McNeill Sr. established the SR Ranch in 1882. McNeill bought eight thousand acres of land granted to the Houston and Great Northern railroad, while on the same date the Kentucky operation bought the sections adjacent. The Kentucky Company offered to buy McNeill out, but he refused all offers. An agreement between the two ranches then allowed both to occupy the same land. Some sort of verbal agreement also called for the Two-Buckle to assume management of McNeill's cattle so that he could stay in Brazoria County and not relocate to the South Plains.[24]

The agreement between the two operations stayed in place for the next six years, but in 1888, following a series of problems in communication and unclear guidelines concerning the contract, relations between SR and the Two-Buckle soured. The Kentucky Company then began a campaign to force the McNeills off the range. The larger company began to force McNeill's herd into a restricted range. The Kentucky operation completely surrounded its smaller neighbor, but in the process of restricting the McNeill holdings, the Kentucky Company also restricted its own. James McNeill Jr. moved to Crosby County in 1902 and took over operation of the ranch. The SR Ranch then entered a period of growth and success. Portions of the SR Ranch still remain in operation as one of the few pioneer operations maintaining a presence in Blanco Canyon.[25]

Hank Smith's closest neighbors were also his relatives. Mrs. Emma Leonard, Smith's niece from Ohio, traveled to visit her uncle in 1887 and became so enamored with Blanco Canyon that she decided to remain in Texas. She met and married storeowner Van Leonard mere months after coming to Texas (in fact, he was probably a large reason she decided to stay), and they were married in November 1887. The couple built a dugout house only a quarter mile from the Rock House, where they remained until 1888, when they moved to the Sandhill community in Floyd County. Emma was not the last of Smith's nieces to move to Texas. Her sister, Mary, also came for a visit and met and married Fred Horsbrough, the manager of the Espuela Ranch.[26]

But it was 1879 that saw the beginning of the largest permanent settlement in the region with the establishment of a Quaker colony twenty miles west of Smith's ranch. Paris Cox, a Quaker and sawmill owner in Indiana, had the idea of establishing a farm community for himself and his fellow Quakers. He had bought a section of railroad land in 1877 and began to gather colonists for his project. Cox's idea was to isolate his colony and allow the Quakers to practice their religion free from the hostility and scorn he had experienced in Indiana. Cox visited the area in late 1878, and one of his first stops was to talk with Hank Smith. Although he did not record the conversation, given his obvious powers of persuasion and his desire to see communities, farms, and families in the area, Smith most likely painted a rosy picture of life in Blanco Canyon. During the visit, Smith, always on the lookout for a business proposition, accepted Cox's contract offer to dig the prospective community a well and to plow and plant crops on thirty acres of land in anticipation of the Quakers' arrival. Smith and Hawse immediately began to dig the well. This was a risky proposition; no white settlers had ever attempted settlement above the Caprock, primarily because there was not a reliable water source. Smith and Hawse were also dubious; Smith went so far as to advise Cox to find land south if he could not strike water. But the well struck water at ninety feet and provided enough water for the farm community for the next five years. Later in 1879 Smith helped the viability of Estacado further when he and Charley Hawse built a road that connected the Quaker community with Hank's ranch at Mount Blanco.[27]

Cox returned to Indiana and began to gather colonists. Eventually, the Quaker group pooled enough money to buy twenty more sections adjacent to the original plot. In the fall of 1879 the first families traveled to Texas and began settlement. The settlement was initially named Marietta (sometimes spelled Maryetta) for Cox's first wife but would eventually become Estacado. Quakers and other potential settlers came to Estacado in the next five years, and in 1884 the town acquired a post office. The Quakers valued education and, by 1882, had established a school and, in 1890, began the first college on the South Plains, the Central Plains Academy, which operated for two years. By the late 1880s, Estacado boasted a population near two hundred; the Llano Estacado now had a town.[28]

With the influx of scattered populations of farmers, ranchers, and the occasional buffalo hunter, the Blanco Canyon area was becoming a crossroads as routes traversed through the region on the way to Santa Fe and points west. But the nearest post office was still in Fort Griffin, over 150 miles to the east. Hank Smith set out to remedy the situation when he began to campaign for a post office in the region. He collected signatures from neighbors as well as from store customers, and sent a letter to Washington, DC, asking for an extension of the Fort Worth–Fort Griffin mail route through the area and on to Fort Sumner, New Mexico. Again, Smith probably had an ulterior motive; the most prominent structure in the region was his Rock House, and if the postmaster general granted his request it was reasonable to assume that the new post office would be located in his home and his store, increasing his business and contacts for more business. And a new post office needs a postmaster, a job that Elizabeth Smith was ready to fulfill.[29]

Any application made to the postal service was required to include a name for the new post office. In Texas, it was common for town names to emerge from the original naming of area. The most prominent feature surrounding Smith's Blanco Canyon home was a stark white caliche cliff just south of Smith's home, known as Mount Blanco. Smith decided to use the name of the feature on his postal application, and when Postmaster General David M. Key approved his request in September 1879, Mount Blanco became the name of the new post office. Soon a little community grew around the Rock House and post office. Elizabeth Smith was also named the postmistress, applying for the job "because there was no one else." She and Hank traveled back to Fort Griffin in October, secured a bond through Frank Conrad, and began operations in late October. Mail ran from Fort Griffin twice a week and dropped letters and packages for ranches and settlements in a radius of greater than one hundred miles.[30]

Elizabeth set up her postal operation in the living room of her home. Most of her customers came from the surrounding ranches, and they usually sent one of their hands to Mount Blanco for the mail once a week. Eventually, Hank constructed a small building adjacent to the home for his wife; it served as post office, store, and the occasional dining room when travelers passed through. Evidently, Elizabeth never let a cowboy, ranch hand, or any other visitor leave on an empty stomach. Elizabeth Smith served as Mount Blanco postmistress until 1916, at the time the longest tenure of any postmaster or postmistress in the nation.[31]

Although the Quaker community of Estacado became a farming endeavor, the primary economic activity on the growing South Plains was ranching. The six hundred head that Hank Smith assumed as part of Tasker's debt were the first cattle in the region, but by the early 1880s other operations had followed Smith. The operations in and around Blanco Canyon were free-range outfits.

The open-range cattle industry emerged after the Civil War when South Texas cattlemen began gathering and trailing cattle to railheads in Missouri and Kansas. This "Texas system," as Terry Jordan has termed it, evolved on the coastal prairies of southwestern Louisiana and Texas with Spanish, Mexican, and southern American roots. It allowed cattlemen to raise thousands of head of cattle while owning only small portions of land. Cattle roamed across the broad expanses of the plains, unencumbered by fences or, in many cases, natural barriers. Ranchers on the plains tended to locate their headquarters wherever the most water was available and ran their stock on public lands where possible, although by the 1870s they began to acquire as many sections as they could around the headquarters. Often, cattlemen simply built corrals and outbuildings on unclaimed portions of public lands to discourage other settlers and herds.[32]

The operation of an American western cattle ranch—although largely a matter of economics, of business, of connecting cow with consumer—is most often associated in the popular mind with a single heroic type, the cowboy. Tall in the saddle, fast on the draw, booted and spurred, independent and free as a prairie wind, the notion of the cowboy as a national hero has become lodged into the nation's memory, so encased in clichés that neither truth or even parody can reduce the reverence and dim the delight that he generates within an admiring public.[33]

Such a description, of course, is closer to myth than reality. It ignores the industry that nurtured the hero; it fails to distinguish, or even identify, who was the most significant figure in the western cattle industry, the range entrepreneur, or, as he is more often referred to, the cattleman. The cowboy stereotype stresses the romantic and individualistic features of western range society at the expense of its cooperative and often corporate character. Most of all, the cattleman was a businessman engaged in a particular industry, the raising of beef cattle. He did not concern himself with looking a part or playing a role. His main preoccupation was in making money. He was above all else a capitalist who pursued ranching as a purely economic activity.[34]

Hank Smith approached his stock operations as just such a capitalistic enterprise. Originally, he envisioned running a large, open-range cattle ranch. The six hundred head he originally acquired had grown to almost one thousand head by 1881. But Smith had little experience as a cattleman; he had spent most of his adult life as a wagon master, a miner, and an expressman. He had contact with cattle operations in Fort Griffin, but only as a retailer or acquaintance; he had never operated a stock business.[35]

But as the operation matured, Smith seemed to have an instinctive idea about how to manage his herd on the precarious Texas South Plains. Chiefly, Smith kept his herd relatively small, particularly compared to his large neighbors. Even many of his contemporaries wondered why Smith was reluctant to expand his land or his cattle holdings. He did not even attempt to secure enough cattle to fully stock his own land. Smith surmised that many of his neighbors were overstocking their range. Some of his contemporaries speculated that perhaps Smith did not desire to become a large landowner or a "cattle baron," that his interests lay elsewhere.[36]

Such an assessment may be partially true, but in Smith's writing and other papers another much less prosaic motive appears that may be closer to the truth. The cattle industry grew in importance, and competition for ranges grew keener as cattle ranges spread to the west. At this point, ranchmen who had the capital soon realized the importance of securing a more permanent source of grazing. This task was made much easier by the relatively liberal land laws of the state. During the Texas Republic, large grants of land were dispersed by the government to endow the state university; as well, 17,712 acres of land were given to each county for public schools. After Texas became a state it continued such land disposal policies, and in 1858 Texas passed a law that granted railroads sixteen alternating sections for each mile of track laid. In 1876 Texas passed a new constitution that called for the alternate sections of tracts granted to the railroads, along with one-half of the public domain, should be set aside in a permanent school fund. But the notoriously frugal late nineteenth-century Texas legislature soon passed further legislation that provided for the sale of these school lands.[37]

The low price of the school lands and the liberal sale terms prompted many of the well-funded cattlemen to secure permanent control of the ranges as competition became stronger. Large cattle operations purchased all the land that they could legally buy and often evaded the legal strictures through buying land in the names of employees or various members of the owner's family. Such a process was repeated in the area of Hank Smith's ranch. Smith also hoped to buy additional land but lacked the funding, and soon after arriving in the Blanco Canyon region, he quickly found his lands surrounded by the larger operations of the region, such as the Kentucky Cattle Raising Company's Two-Buckle. Soon after its establishment, the Kentucky investors began leasing and buying any land available in the area of its holdings. Smith and the smaller landholders were at a disadvantage in the situation; they could not hope to match the capital behind operations such as the Kentucky Cattle Raising Company. Eventually, operations such as the Two-Buckle collapsed under the weight of drought, severe winters, and a depressed cattle market, but Smith and many of the other smaller cattle operations survived, either through luck or, perhaps, design.[38]

Other than the original tract that he foreclosed from Charles Tasker, Smith, in 1882, purchased 320 acres of Eastland County School Lands. He purchased no more land until 1897, other than a forced purchase in 1890. During that year, the surveyor at Estacado discovered that Smith did not have clear title to his homestead. Charles Tasker had optioned the land in 1877 but never completed the purchase. Thus, for Smith to secure his title, he had to buy the 120-acre plot that surrounded the Rock House. In 1897, he bought six sections of railroad land adjacent to his acres, bringing his total land holdings to ten sections. On the semiarid South Plains, it was a small ranch operation, and Hank Smith never used the entire acreage, preferring to lease a large portion to other ranchers and farmers.[39]

The cattle that Charles Tasker originally brought to Blanco Canyon were probably Texas longhorn crosses. During the period, the notion of pureblood stock on the open range was not usually considered an option. When cattle had to be pastured exclusively on the open range, operators did not stock purebred or high-grade cattle since they could not be sure of having exclusive use of these expensive animals. Smith was a committed open-range rancher and in the first few years of his operation had almost the exclusive use of the Blanco Canyon range. Smith and Hawse did most of the cattle work alone, although they occasionally hired temporary workers to help with roundups and branding, usually members of buffalo hunting crews that by the early 1880s had found most of the buffalo gone from the South Plains. Hawse had probably worked cattle in the Fort Griffin area and was very comfortable, indeed adamant about, tending the herd from horseback. But Smith, although an able horseman, was not as traditional as Hawse. His only previous experience with cattle was the short time he had spent working on Abe Sutton's small ranch in the Gila River, and he had done most of his work from the ground. He seldom engaged in cutting or roping cattle and was much more interested in his sheep. George Smith was the true cattleman in the Smith family and regularly accompanied his father and, more frequently, Charlie Hawse, out to tend the herd.[40]

One aspect of the cattle operation that Smith insisted on making most of the decisions about was the sale of any animals. Smith sold most of his calves and bulls to the larger operations in the area, most regularly to H. H. Campbell and the Matador Land and Cattle Company and the Espuela operation, particularly after his in-law Fred Horsbrough took over as manager at Espuela. He also sold to the SR and TM ranches. He was a keen bargainer and Campbell constantly complained about Smith "getting the better of him." Horsbrough remembers his relative as "stubborn as a stump when it came to pricing cattle." Although he occasionally did work for the Two-Buckle, there are no records that indicate Smith sold cattle to the outfit. Although he did contract some of his stock to trailers to take to railheads for shipment, Smith's cattle operation, until the 1890s when George took it over, remained very small.[41]

Sometime in the 1880s, Smith adopted a brand for his ranch, the +B or "Cross B," probably named after the county. He continued to run his ranch as a small operator well into the 1890s, selling a few cows and steer each year but nowhere near the number of his neighbors. He also continued an open-range operation, long after most of his fellow ranchers had abandoned the method.[42]

George Smith had other plans. He was the cattleman of the family and in 1895 Hank Smith, by then sixty years old, turned the cattle operation over to George, although he did occasionally give his son advice on the sale of cattle, much to George's consternation. George Smith wanted to build a pureblood stock operation, and he believed that the best cattle for the South Plains was the Hereford. Originally an English breed, Herefords became a favorite stock for many Texas cattleman due to their hardiness and tendency to "fatten early," producing a tender cut of beef. The Herefords that producers brought to Texas generally came from the midwestern Corn Belt; many ranchers began to use them exclusively, although they tended to also have a strain of Shorthorn blood. George Smith was one who began to stock the +B exclusively with Herefords. He began to sell the ranch's other range cattle in 1895 and, by 1900, the Smith operation consisted solely of Hereford cattle.[43]

Hank Smith survived on the South Plains probably through a combination of luck and design. When he arrived in Blanco Canyon it was with the express idea of operating a cattle ranch. But, as larger operations moved into the area and the grazing land became more restricted, he was forced to alter his plans—he began to stock his range with sheep as well as cattle. If measured by pure numbers, he was much more a sheepman than a cattle rancher. Simply to survive, by 1884 he had purchased five thousand head of sheep. Combination sheep-cattle operations reduced the financial risks often associated with sole animal stock operations. The same conditions that made the South Plains practical for cattle also made it suitable for raising sheep. The Blanco Canyon region had abundant grasses and a good water source. As far as the record indicates, Smith had no practical experience in sheep raising before moving to the South Plains. But as an expressman, he had traveled extensively in Texas during the late 1860s and early 1870s, years that coincided with the Texas sheep boom. He no doubt observed sheep operations in the central Texas region during his travels to San Antonio, as well as some scattered sheep flocks in southern New Mexico and the El Paso area.[44]

The Texas sheep boom began with the end of the Mexican-American War when new arrivals of stockmen in South Texas began to stock their land with sheep. Although the Rio Grande Plain proved viable as sheep country, the industry stayed fairly stagnant through the Civil War. Numerous factors caused the sheep industry to boom after the Civil War. Changing fashions in the 1850s meant wool became favored over cotton, so when the Civil War cut the supply of cotton to New England textile mills, the capitalist merchants simply switched from cotton to wool. Also, cotton was associated with the Confederacy, rebellion, and slavery, so patriotism caused many Northerners to switch to wool fabrics. Wool processing techniques improved and wool began to truly rival, and in some areas surpass, cotton as the fabric of choice.[45]

With demand rising, wool prices also soared, further contributing to the boom. As the war ended and the return of the US Army to the frontier forts expanded the frontier line within the state, thousands of people moved into the western reaches of Texas. These settlers were primarily farmers, but they also became stock raisers. Since sheep were now a good investment, many began to run the animals on the range. Eventually, with the open land in western Texas available for grazing, flocks in the region began to number in the thousands, sometimes as high as ten thousand. The number of sheep in Texas rose to more than five million by 1887.[46]

With the rise in demand and prices, Smith's sheep operations became prosperous. Smith's wool clip for his sheep was productive for the dry South Plains. During the 1870s and 1880s the Texas average for wool per sheep was between five and seven pounds. The better-fed and heavier animals in the central and eastern areas of the state sometimes produced wool clips of as much as twenty pounds, although this was obviously an anomaly. Smith sheered his sheep once a year, every May. It was a task that he obviously enjoyed because he insisted on personally clipping most of the sheep, although Charlie Hawse and his two sons were more than capable. According to his records, from 1885 through 1897, he averaged four and a half pounds of wool per sheep. He averaged a flock of approximately seven thousand during those years, and his wool sales became quite profitable. In his peak year, 1890, he reported selling over 37,000 pounds of wool for a total of $13,433.37.[47]

Smith took a keen interest in the raising of his sheep in the first years he stocked his land, but he eventually began to turn the sheep operations over to his two sons, Robert and George. In 1892 he bought five thousand more sheep, which brought his total to over ten thousand. He decided that this was far too many for Blanco Canyon to sustain during the winter, so he divided his herd and sent his sons to drive the flock to winter grazing ground in Stonewall County. Eighteen-year-old George and thirteen-year-old Robert left soon after Christmas and moved along the Salt Fork of the Brazos to Double Mountain and on to the winter grounds.[48]

The two Smith sons were to remain with the sheep through the winter, living out of a camp wagon. Smith entrusting his sons with half his sheep herd was proof of the degree of trust he had in his sons and also how much these two young boys had learned since moving to Blanco Canyon. The winter obviously went well for the sheep as they returned well fed; they yielded more wool than the flock that had stayed on the South Plains. However, George Smith had to return early in late February when he fell ill, leaving thirteen-year-old Robert alone. George rode over one hundred miles back to Blanco Canyon, covering the distance in four days—a remarkable pace for a sick young man. Along the way his illness worsened into pneumonia, and for over a week he hovered near death. After convalescing for a month he then rode back to the flock and helped Robert drive the sheep home. Smith felt that his sons had proven their worth as stock hands, and over the next three years they repeated their journey. Perhaps his bout with pneumonia soured George on sheep raising because after 1895 he took over the cattle operations of the ranch and Robert concentrated on the sheep.[49]

The presence of sheep on the South Plains range tended to anger some of the larger, cattle operations. The ranch hands and operators of the Kentucky Cattle Raising Company's Two-Buckle Ranch held sheep and sheepmen in complete contempt. But they also scorned the small cattle operator and the farmers at Estacado, so their rancor was probably a continuation of their attempts to secure the majority of the range for their operation. Texas cattlemen steeped in the tradition of the cattle culture often considered sheep smelly and silly; they left the production of sheep to the people that they deemed worthy of such ignoble efforts—Mexicans, Germans, Basques, and Englishmen.

There is no doubt that many of these notions remained with some of Smith's neighbors. But, raising sheep had become a profitable venture and one that could be the difference between profit and loss for the small stock operator. Thus, many of Smith's smaller neighbors, such as the McNeil SR Ranch and the TM operation also ran sheep. Most of the cattlemen on the South Plains accepted sheep as an economic necessity, and the stereotype of a cattle versus sheep range war is overdrawn, at least in the Blanco Canyon region.[50]

Besides beginning to focus on sheep, Smith had other motivations that kept him from expanding his land holdings and his cattle stock. Hank Smith had quite a few business experiences in the years since he had left Ohio. He had seen his and other businesses fail due to over- or too-quick expansion, and he probably understood that his Blanco Canyon venture was his last opportunity for some semblance of prosperity. A small herd and operation was a manageable herd, one that he, Charley Hawse, and eventually his sons could work alone. If one of the many natural or market catastrophes that so often occurred within the volatile agricultural market struck his holdings, he could continue operations and recover sufficiently enough to continue his business. This German immigrant, a man of little education but obviously high intellect, seemed to grasp the nuances of a market economy better than many of his neighbors. Smith, like the vast majority of westering people, understood that life in the West was no romantic adventure, and agriculture was no refuge from a commercial market. Operating a stock ranch or a farm is to be intricately intertwined in a market economy. Smith, and others like him, came West to build a financial future, not to begin an entirely different culture based on a specific and unique regional character.[51]

Hank Smith was more an entrepreneur than a stockman. Throughout his life his pattern had been to search constantly for expanded economic opportunity. After he had plowed and planted the first agricultural crop on the South Plains for the Quakers at Estacado he became convinced that the South Plains could be exploited as a crop growing region. In this he was prescient; in the twentieth century the area became a center of agricultural growth and today remains one of the largest agricultural regions in the country. The year after planting the Estacado crop, Smith and Hawse cultivated land in Blanco Canyon. On thirty acres of flat land near the Rock House, Smith planted a vegetable crop. In the following years he experimented with many kinds of vegetable and grain crops and reported that the region produced favorable yields of most crops. He also did not end with farm produce. When he found a wide variety of wild plums, grapes, and currants growing in the canyon, he became convinced that West Texas could become the center of a fruit industry. To prove his point, and always in an attempt to find another source of revenue, he planted a wide range of fruit trees, including peaches, plums, and apples. He constantly encouraged new arrivals in the region to plant fruit trees and experiment with different crops. Right up until the time Smith suffered a stroke in 1910, he could be found working in his fields and was always open to sowing new crops.[52]

In the 1890s, Smith began to lease land to tenant farmers. Included in the lease contracts were his ideas about diverse use of the land and the need to plant a variety of crops. His standard agreement was a lease of seventy-five acres, and Smith provided all implements, wagons, and horses necessary to cultivate the land. Tenants had to agree to plant a combination of cotton, kaffir corn (a forage crop), corn, and milo maize. The tenant had to cultivate the entire lease, and cotton could not exceed twenty acres. Smith was to receive one-half the crop in payment, with the grain delivered in the field and the cotton after ginning. Smith agreed to pay half the ginning expenses and to provide the tenant a residence and board in exchange for "chores."[53]

Because George and Robert Smith increased their roles in the management of the +B Ranch after 1900, Hank was free to pursue other business interests. He had gained the same reputation in Crosby County he had enjoyed in Fort Griffin, that of a man who was always interested in hearing and considering a business proposal. James Posey, a former cashier of the First National Bank of Floydada, approached Smith in 1906 to inquire about his interest in investing in a new venture, the Plainview Bank and Trust Company. Smith listened and was obviously intrigued. Although he told Posey that he was willing to invest $1,000, Posey's initial offering called for investments of $2,500. Smith then declined Posey's proposition, explaining, "Speculation in land and stock is risky enough," but he urged Posey to "keep him in mind if he heard of any other ventures."⁵⁴

After 1900 Smith renewed an interest in the mining claims he had made in the Gila Mountains before the Civil War. Smith tried to revive his claims in 1908 and was still convinced that the Gila Mountain claim that he had prospected might still yield gold. At the age of seventy-two, he began corresponding with his nephew Joseph Stercle (?), who had traveled to Arizona to work in a lead smelter but had turned to gold prospecting. Smith related the approximate location of his old claim and then offered to join him in Arizona. He began organizing an expedition and apparently recruited local volunteers to accompany him to the Gila River. But when he wrote the US Department of the Interior to inquire about the legality of his claim, he was informed that there was no claim in the specific area in his name or any other. Smith vociferously disputed Interior's documentation and attempted to establish a legal claim in the area. Although the Interior Department repeatedly denied his requests, he decided to travel to Arizona anyway and resume his mining career. But before he could leave Blanco Canyon, he suffered a stroke, ending his last case of gold fever.⁵⁵

During the same period as his attempt to resurrect his gold claims, 1904–7, Smith also began to try to revive another claim he had made in his earlier life. When he was operating his freight business in Fort Quitman, he had filed a loss claim with the US government concerning the theft of mules and forage during a Mescalero raid in 1870, a claim that was never resolved. Financially secure and with time on his hands, he began efforts to recover his earlier loss. To press his claim, he retained Washington, DC, attorney John Wharton Clark, who advised Smith that, at present, he could not recover his loss. The lawyer explained that because Smith's loss was due to the fact that the government considered the Mescalero hostile until spring 1871, the Amity Clause of the Indian Depredation Act kicked in and stipulated that property could not be recovered unless the Indians responsible were under the auspices of the federal government at the time of the loss. However, Clark was optimistic that Congress would strike the Amity Clause in the next session, and he would move forward on the case at that time.⁵⁶

Clark pressed Smith's claim before Congress for the next two years. He continued to express optimism about Smith's chances for recovery, but Congress denied all attempts at recovery. Clark stopped sending updates to Smith in late 1906 concerning the matter, causing Smith to send him a scalding letter inquiring as to why the lawyer had ceased correspondence. Clark replied that he was still working on the matter and would continue during the next session. But Clark wanted a larger retainer, another three hundred dollars (Smith had already paid him two hundred dollars). Smith forwarded no additional money, and in October 1907, he told Clark to drop the matter entirely.[57]

Smith's entrepreneurial activities on the South Plains of Texas were the culmination of a pattern that he began to form the moment he left Ohio—everywhere he settled and every activity that he undertook was with one overarching purpose: to make a profit. Daniel Webster once wrote of the "growing propensity" of Americans to seek wealth as they pushed westward, and Washington Irving coined the term "Almighty Dollar" when he related the experiences of fur traders in the West and their search for wealth. Both men surmised that it was the West itself that instilled the desire for the accumulation of wealth in migrants to the region. Others have associated this urge for riches and opportunity as the primary factor in the development of the individual that is supposedly indicative of westerners. But the more accurate depiction would be that westerners were simply following a pattern that had long characterized Americans as a whole. The US Constitution was more than anything else a blueprint for an economic way of life, the creed of self-advancement that manifests itself in economic progress. Americans, perhaps more than any people before them, became convinced that wealth and material possession were essential for happiness. Hank Smith, in this regard, truly became an American and a "westering man." He was a practical businessman whose focus was making sure his businesses were profitable enterprises. In pursuing profits and participating in a capitalist market system, he was not unique because of where he lived or what he did for a living, but was a man of his era. In seeking greater profits or exploring new markets and activities he was no different from the much better known capitalists of his era, such as

John D. Rockefeller or J. P. Morgan. While he did not earn their great wealth, he did have the same motivations and eventually lived comfortably. This westering man's life was grounded in economic reality and the search for profits amid plenty of frontier competition.[58]

7

Crosby County's Most Prominent Citizen

From the moment he arrived in Blanco Canyon and began his +B Ranch, Hank Smith promoted the region as an ideal locale for settlement, a tradition that he shared with other westerners of his era. He wrote ebullient letters to the *Fort Griffin Echo* inviting other settlers to sample Blanco Canyon's abundant game and seemingly endless cattle grasslands. He pored over government topographical reports on the South Plains and took great pains to point out the resources of the area, most importantly the availability of water. During the late nineteenth century, Blanco Canyon had four springs, of which the largest is the source of White River. He claimed that there were great underground water sources "that no bottom had even been found for any of them." He also reported the existence of a vast artesian spring on his property and often cited an army report by a Captain Livermore that described an "inexhaustible supply of water."[1]

In his efforts to promote settlement and agriculture on the South Plains, Smith, in 1886, took on the task of weather recorder for the United States Weather Bureau and continued the activity until 1907. He meticulously recorded the temperatures, precipitation, and speed and direction of the winds. While he received a small salary, Smith's motivation in the undertaking was more informational, part of his overall campaign to promote the region for settlement. The records indicate the great variety in the amount of rainfall on the South Plains, a constant irritant for the stockmen and farmers of the region. While the average rainfall from 1886 through 1907 was just over twenty-two inches, that average was skewed by years of abnormally high rainfall and years of much lower totals. For example, the Blanco Canyon area received almost thirty inches of rain in 1900 and over thirty-two inches in 1905. But in the years between those, the average rainfall was only approximately sixteen inches. Such variances made the region a precarious location to raise crops and stock.[2]

Smith's activities in promoting the region are part of a larger pattern that was characteristic of other places in the American West. From the years immediately after the Civil War through the early twentieth century, western boosters, in the words of author David Wrobel, "literally tried to imagine western places into existence through embellishment and effusive descriptions." Western boosters were overwhelmingly optimistic men who told potential settlers exactly what they wanted to hear about the West, even if that meant exaggeration. They portrayed the region's future as exceedingly bright and glossed over any pitfalls that a potential resident might face. They often presented desolate frontiers as rapidly populating areas devoid of risk or any potential for failure. Western regions were presented as virtual Gardens of Eden and places of endless opportunity.[3]

Like Smith, some western boosters were themselves recent arrivals to their region, and they enthusiastically and imaginatively portrayed their western homes in glowing terms because they wanted their own dreams and businesses to succeed. Such promoters were often also selling dreams to sell products. Smith no doubt was keenly aware that more residents in the Blanco Canyon region meant more opportunities and more profits for his own business ventures. But regional boosters were more than just earlier versions of today's advertising executives. Many of them were optimistically looking to the future, describing what the region *could* become instead of what it currently was. The West Texas city of Abilene, founded in 1881, described itself as the "future great city." Thus, boosterism often confused past and present.[4]

But to explain away the often-erroneous depictions of a region's attributes as an optimistic forecast of the future is itself superficial. Promotion of the American West in the nineteenth century was more than just innocent bragging and exaggeration of the details. Some of these boosters were charlatans looking to increase their profits by duping unsuspecting potential customers. While Smith's promotional letters and depictions were not outright lies, and his motives did seem at least outwardly pure, the letters sometimes fell into the category of overly optimistic, and he did stand to profit from a more populated area.[5]

Whether or not Smith's promotional activities had a direct effect, the region, after 1880, began to gain population. By 1882, the Quaker settlement of Estacado could rightly be called a town; the community gained a post office in 1881 and in 1882 the first school on the South Plains when Emma Hunt began teaching in a dugout classroom. As more families moved to the community, the school soon outgrew the small dugout and moved into the Quaker meetinghouse. By 1890, Estacado had grown to a population of over two hundred. George and Leila Smith attended the Estacado School in the mid-1880s, probably boarding with local families, but Elizabeth decided, by 1889, to tutor her children at home. The town soon boasted a newspaper and numerous retail businesses. Although it had begun as a Quaker settlement, by 1890, non-Quakers outnumbered members of the Society of Friends.[6]

The area around the Smith's home had also grown into a community, Mount Blanco. The 1880 census listed seven people as residents of Mount Blanco, but, surprisingly, no members of the Smith family were counted. Mount Blanco had grown to thirty-seven residents in the 1890 census. When Elizabeth Smith brought her children home from the school in Estacado she also took the first step in organizing the Mount Blanco School. Other children soon joined the Smith children at the Rock House. As the number of children grew, Hank and Charlie Hawse built a small schoolhouse one-quarter of a mile west of the Rock House. Elizabeth served as the first teacher, but by 1893 the community hired permanent teachers, two older Quaker girls from the Estacado School.[7]

The Texas legislature had designated Crosby County in 1876 with legislation that created the Crosby County Land District. It was drawn from existing Young and Bexar Counties and was named for Stephen F. Crosby, a three-term commissioner of the General Land

Office. The huge tract also included the present counties of Lubbock, Hale, Floyd, Lamb, Bailey, Hockley, Cochran, Dickens, and Motley. Although Crosby County existed as a legal entity, throughout much of the 1880s it remained unorganized and was attached to Baylor County for judicial purposes. Smith maintained that the large cattlemen in the area, principally the Two-Buckle men, were dead set against organization and for a decade had blocked all efforts to form a new county. The cattlemen did not want an accurate count of their cattle or land holdings for tax purposes and preferred the isolation of ranches far from the seat of government.[8]

As new residents moved to the area and Estacado grew, the dynamics began to change. Thus, as the 1880s began, several residents began to discuss organizing Crosby County into a political entity. At the same time, the Texas General Land Office sent out "grass commissioners," men whose task was to force the stockmen using the rangelands located on school land to pay rent for the privilege. George Swink traveled to Crosby County in 1883, and for two years he contacted stockmen and attempted to collect the proper rent. Most of the herd owners refused to pay, and, probably through political pressure, Swink was dismissed as grass commissioner for the area. Frustrated and no doubt seeking some measure of retribution, Swink began to campaign in Estacado for county organization.[9]

Many of the Quakers reacted favorably to Swink's proposal, but it took at least 150 signatures of registered and eligible voters to sign the petition to bring the matter to a vote. Swink was determined; after all, he was out of work and hoped to find employment within the new county's government. He drew up the petition, but to get the necessary signatures he asked Hank Smith to travel the huge tract to gather them. Smith, who was in favor of organization, agreed and began to journey throughout the Crosby Land District attempting to persuade the residents to commit to organizing Crosby County. Smith reported that most of the signers were Quakers and the few buffalo hunters who remained, but most of the "cowmen flatly refused to have anything to do with it." Smith, forever the businessman, began to worry how the county would reimburse him for charges if the campaign and vote failed. At a meeting in Estacado, after the petition had the required 150 signatures, Smith was assured that he would be paid for all services if the supporters succeeded in organizing Crosby County.[10]

Smith next took the petition to Seymour, the county seat of Baylor County, a 130-mile trip. The Baylor County commissioners had motives for creating Crosby County as well. Prior to 1886, all the territory from west of Baylor County to the New Mexico line was attached to Baylor County for judicial purposes. Knox County, adjoining Baylor County to the west, had organized earlier in 1886. As the legislature convened in January 1887, it was likely that all the unorganized territory would then be attached to Knox County, causing the tax revenue of the area, which included several large ranches, to flow into the Knox County treasury. So, in an attempt to keep its rival county from gaining revenue, the Baylor County commissioners approved the petition and set an election.[11]

The organization of the county required the election of county officials. On September 11, 1886, George Swink was elected county judge; Paris Cox, clerk; Felix Franklin, sheriff; and Hank Smith, tax assessor. All of the new officers, other than Smith, were residents of Estacado, which was also designated as the new county seat. The officers next had to stand in the general election in November; the voters retained all (most ran uncontested) except for Smith. In the general election, T. A. Gray challenged Smith and they each received twenty-nine votes. For some unknown reason, Gray did not run in the special election to decide the office, and Smith beat two candidates for the position. The Texas legislature approved the organization and election during its 1887 session and also organized the Thirty-Second Judicial District, comprised of twenty-five unorganized counties attached to Crosby County for election and judicial purposes.[12]

With the county organized and a county seat established, the first order of business for the new county was to build a courthouse. Estimates, even for the modest building that Crosby County hoped to build, showed that it would take a minimum of $8,000 to construct. New Crosby County immediately faced revenue problems. Most of the county's tax base was in the form of rangeland and cattle. The large cattle corporations, who would be the largest taxpayers, paid their taxes directly in Austin and most often undervalued their stock and holdings. The rest of the assessed county value, outside of the large ranch sources, was not sufficient to raise enough money for administration and building a courthouse. The state comptroller advised the commissioners that the cattle companies were now subject to local taxation, which required an accurate assessment, and that fell within Smith's purview as tax assessor. An accurate count of their stock, improvements, and holdings for tax purposes was the last thing that the large cattle companies wanted and was the main reason they opposed any county organization.[13]

Perhaps the most affected by an accurate assessment would be the large Matador Land and Cattle Company. Since its sale to a Scottish syndicate, ranch manager H. H. Campbell, probably under strict orders, had underreported its stock holdings for tax purposes. It also refused to pay for grazing rights on school land and certainly did not want a public record of exactly how much public land they used for ranch operations. Other large stock operations in the area, such as the Two-Buckle and Espuela, were also resistant to an assessment and an accurate accounting of the amount of public land they used.[14]

Although he faced opposition from the large ranch owners, or perhaps because of such opposition, Smith began a meticulous assessment for Crosby County. In many ways, an accurate count and the resulting taxes represented a way for the small landholders to finally wrest some power and influence from the large operations. The opposition certainly did not intimidate Smith. He began to travel the massive new county, one that covered what is today a ten-county area and encompassed most of what is referred to as the South Plains. He did most of his traveling in the spring and summer and visited every ranch headquarters, farmhouse, and merchant in the area. His primary responsibility was to make sure that the livestock count the owners gave him was accurate and assess resident landholdings. He made his rounds in a two-horse wagon and carried a partially used ledger book left over from his time as a clerk in the Conrad and Rath store in Fort Griffin. He took copious notes in the margins and described each ranch, its land improvements, and the number of stock that each owner claimed and then he followed this count with his estimate from his survey. For example, the Matador had estimated 38,000 head for tax purposes, but Smith's assessment cited the Matador as having 64,000 head. The actual size of the herd, as reported to the Matador headquarters in Dundee, Scotland, was over 96,000.[15]

As expected, the larger operations in the county disputed Smith's assessment and contested his estimates. No operation raised as many objections as the Matador. Its leaders objected to Smith's records, but when the county government in Estacado, made up entirely of small farmers, ranchers, and merchants from the county seat, certified Smith's assessments and levied taxes based on his records, the Matador decided to contest Smith politically. Smith stood for reelection in 1888, and the Matador recruited an employee of the ranch, Joe Brown, to oppose him. The Matador managers ordered all their hands to vote for Brown, and they were successful in gaining the cooperation of other large operations in the county. Brown defeated Smith by a small margin, and Smith lost his job as tax assessor for Crosby County. Smith remained bitter about the defeat and heightened his dislike for the "corporations" that defeated him.[16]

The late 1880s and early 1890s were not kind to the county seat of Estacado. Drought, prairie fires, insect infestations, and internecine squabbling began to take a toll on the town. The precarious nature of South Plains agriculture perhaps took the biggest toll. Although the early years of the Quaker colony were blessed with adequate and sometimes abundant rainfall, the mid- to late 1880s brought a serious and catastrophic drought for the farm community. The drought was interspersed with a grasshopper invasion that devastated the crops. Added to such problems was that, while the Quakers had hoped to build an isolated and homogenous religious colony, as the town grew, nonQuakers began to outnumber the original Quaker settlers. The non-Quaker settlers began to agitate for their own religious preferences, and they objected to the parochial Quaker school and teachers at the Estacado School. The Quakers seemed to have the stronger position due primarily to the strong leadership of Paris Cox, a man respected in both camps. But in 1888, Cox died and the new Quaker leader, Frank J. Brown, did not have the strong presence of the town's founder. Estacado's primary asset was the county seat, but in 1890 the town received a new rival that threatened to take the county seat.[17]

The discontent over the situation in Estacado led merchants H. E. Hume and R. L. Stringfellow to leave the county seat and buy a section of land southeast of the town and lay out a new townsite. Hume named his town Emma, supposedly after Stringfellow's wife, although the origin is under dispute since records indicate that Stringfellow's wife was not named Emma. Emma Hunt was the first schoolteacher at the Estacado School and could have been the origin of the name, but given the acrimonious nature of the dispute between the two towns, such an origin may also be dubious. Stringfellow and Hume organized a general store, laid out the townsite, and opened up a public sale. Several families then moved from Estacado to Emma, and Stringfellow offered a home and newspaper office to induce J. W. Murray, the owner and editor of the *Crosby County News*, to relocate his influential paper to the new town.[18]

Hume and Stringfellow next petitioned for an election to move the county seat to Emma. The election was held in October 1891 and in a close vote of 109 to 103, the voters chose to move the county government. The courthouse was dismantled and moved at a cost of over $3,000, and Stringfellow and Hume agreed to furnish free water and offices for all county employees until the courthouse was reconstructed. But the two broke that promise quickly and charged an exorbitant rent for the county offices. When the courthouse was finished, they informed the county that the new government would have to pay for any further use of the town well.[19]

In the first years of the new county seat, the town experienced a boom. An influx of people from East Texas eager for land settled in the area as well as most of the non-Quaker residents of Estacado. Smith called Emma a "typical western cowboy town." He apparently held the town in low regard; in notes that he made intended to finish his memoirs he wrote the following description of the new county seat:

> Thousands of cattle being loose on the range, the public well at Emma was a common watering place for thousands of head of stock that did not belong within twenty miles of town. In the bloom of youth Emma was in no sense of the word a virtuous model young town. It had a Tarantula Emporium and a drug store which made two saloons, and for

rotten dope ladled out to mankind Emma was a glittering success. In those days the courthouse was a merry old place and smelled more like the fumes of a booze-factory than a temple of justice.[20]

Hank Smith had become emotionally attached to Estacado, Paris Cox, and the town's Quaker colonizers. He no doubt admired the leader's virtue and the town's resistance to the vices that often accompanied frontier cow towns. The Quakers were the first to join him on the South Plains as permanent settlers, and he and Paris Cox had remained close friends. While Smith, a Catholic, did not ever mention attending the Quaker Church, Elizabeth and the children were frequent visitors, although Elizabeth remained a staunch Presbyterian. Smith was also proud of his role in the organization of Crosby County, and the residents of Estacado were his close allies in that fight. He believed that Emma became another attempt by the large cattle operations in the county to dominate county politics. He claimed that the town's "main draw back" was that it was surrounded and dominated by the "large corporations and individuals" that wanted the range to remain open to only their cattle and the same ones who had originally opposed the settlement of the county. In his mind, it was their attempt to wrest power from the small landholders and farmers in Estacado.[21]

But like many boomtowns, Emma's life was short. Construction began on the Crosbyton South Plains Railway in 1910, and when the right-of-way missed Emma by five miles it rang a death knell for the town. The new town of Crosbyton was on the railroad line, and it began to eclipse Emma in terms of residents and stature. Eventually, in late 1910, Crosbyton won another county seat election and took the seat of government from Emma. Many businesses and residents simply did what they had done twenty years earlier—picked up and moved buildings, implements, and people to the new county seat. Others, such as the post office and a general store, moved to another new Crosby County town, Ralls. Eventually, there was nothing left in Emma, not even weathered buildings that would make it a ghost town. Today, there is no sign that a once-thriving county seat rested on the site, just a state historical marker and the Emma cemetery.²²

Smith did not remain out of the political arena for long. After he lost the tax assessor office in 1888, he did not return to county-wide politics, but instead he set his sights on an office that probably reflected his new ideas on what the county he had helped organize had become. Smith was close to sixty years old, operated a successful stock operation, and decided to remain closer to his Mount Blanco home. He ran for county commissioner in 1892 from precinct two and served a two-year term, which seemed to be the tradition in the county. He also began to turn more and more of the +B operations over to his sons George and Robert, and he became something of an elder statesman of the South Plains.²³

During the 1890s, Smith never ventured far from the Rock House, but he received many visitors. The Rock House became a favorite gathering place for dances, picnics, and other social events. The backdrop of the canyon and the shade of the many trees that Smith had planted made it a pleasant refuge for the many visitors who came to pay homage to the man who had played such a large role in developing the region. The Smiths had carried the monikers "Uncle" and "Aunt" Hank from Fort Griffin to the South Plains, and they delighted in serving as hospitable hosts. One of his favorite activities at these gatherings was to regale an audience with his recollections of his earlier life. Smith cut an impressive figure. He was well over six feet tall, wore a full beard, and, to most, looked the part of a legendary frontiersman. He mesmerized his audience with his tales of his days as a bullwhacker, a miner, fighting the Apaches, and a resident of Fort Griffin, which had already gained a notorious reputation.²⁴

The front porch soliloquies probably began the transformation of Hank Smith from a man who came West to make a living into a regional frontier legend. Like other men from his era, Hank Smith understood his audience and knew what they wanted to hear. The triumphal tales of pioneer men were already dominating the public perception of the West, much as they do today. The people who gathered to hear his stories did not want to know the mundane and often harsh details of what it took to make a life in the West. They wanted to hear the myth, the veil of the pioneering spirit that in their mind made the region different from the rest of the country. Hank tailored his stories to meet their expectations and emphasized what he felt were the romantic aspects of his life, not the actual realities of failed businesses, rootless wandering, or the often harsh and moral ambiguity that accompanied encounters with Native Americans.[25]

Smith became the unofficial historian of the region. Although he had little formal education, he obviously possessed a keen intellect. Easily learning new languages marked his intelligence. As a young German émigré he had learned English quite quickly, within a year of his arrival. He also was able to gain a sizable and workable knowledge of Chiricahua and spoke Spanish quite well. After moving to Blanco Canyon, he became well informed of the history of the region and a great lover of books. His meteorological records of the area were quite extensive, and he was a knowledgeable source on the plant life, agricultural techniques, and the wildlife of the canyon. He urged settlers to grow different and diverse crops and stayed current on the most advanced techniques concerning farming and stock raising.[26]

At some point in the latter years of his life, Smith decided to compile his memoirs. He had always been a voracious note taker, chronicling his life as he moved West. In 1900, he began to write about his life with the ultimate intention of publishing his recollections. Probably influenced by the tales he told to visitors on his front porch, his memoirs tended to emphasize what he considered the romantic aspects of his life and glossed over what were probably the most meaningful and influential portions, those that truly defined him as a westering man. The writings are dominated by stories and tales of Indian battles and the romanticism of searching for lost gold mines. He spent considerable time and effort recounting his exploits as an "Indian fighter" involved in the white settler battles with the Apache forces of Cochise and Mangas Coloradas in the late 1850s. But this period of his life had lasted only about two years and consisted of only one small battle in Pinos Altos. He also took great pains to make his readers and contemporaries aware of the fact that he had met such western "legends" as Judge Roy Bean and Billy the Kid. His memoir seemed influenced by the pulp fiction of the era, the larger-than-life struggles of a pioneer on the western frontier. Perhaps it was Smith's life as he chose to view it, but his life was in reality much more than a series of romantic adventures. Smith knew his audience, and he tailored his tales to meet the expectations of those wishing to hear the life of a frontier pioneer.

Smith's memoirs present somewhat of a problem for modem readers. Smith recounted actual events in his life, but he also tailored his life to cope with a frontier existence. The danger occurs when such expectations begin to become the constructed reality. The presence of the mythic identity of the West, by the late nineteenth and early twentieth century, became an integral part of the history of the region. Smith was no doubt aware of this mythic construction of a unique western identity, either consciously or subconsciously, and it formed a major part of his personality and thus his memories.

The role of memory is a powerful component of a constructed identity, and no region has used the reconstruction of selected memory to a greater advantage than the American West. Richard White calls the "American West . . . the most *strongly* imagined section of the United States." He sees the formation of a western imagined identity as beginning before the conquest of the region was ever completed. As he recounted his life, Smith did not intentionally inflate or exaggerate his adventures, nor did he fabricate his tales. But in many ways his "memories" of his life were patterned to fit the popular perception of the American West, perceptions that he was certainly aware of in the early twentieth century.[27]

When Smith began writing of his earlier life, the actual "Wild West," if it had truly existed, was itself just a memory. But numerous writers, showmen, and chroniclers were already hard at work turning the memory into legend. Wild West shows toured the country, touting the experiences of westerners and the West. Theodore Roosevelt, who had just entered the presidency when Smith began his writing, had skillfully used his image as a "cowboy" to great political advantage. Smith tapped into these same themes and highlighted the same style of adventures when he began recounting his life.[28]

Another facet of western identity, and one also present in Smith's memoirs, was the unique formation of a frontier cult of masculinity. As early as the 1840s, Alexis de Tocqueville found a distinctive layer of personality beneath the exterior of the American male, one of a bristling ego, an entrepreneurial soul, and profound anxiety. Tocqueville's perception would suggest that there was something unique about the notion of American manliness, or masculinity.[29]

The movement west, with its accompaniment of Native American conquest and transformation of a pristine wilderness into productive land, created a unique American masculine manifesto that remains pervasive in American life. The cult of masculinity is particularly present in popular culture and is demonstrated today through film, rodeo, literature, and other venues. While the construction of American masculinity is not a sole product of the American West, the region does represent masculinity's greatest manifestation. For example, when Theodore Roosevelt began his political career, a harsh reality confronted the idealistic politician. His patrician background, sickly childhood, and Victorian concept of manliness, which emphasized self-restraint and sobriety, opened the young twenty-three-year-old to ridicule, deriding his manhood. Roosevelt quickly realized that if his political career was to survive he needed to transform his image. His solution was to become the embodiment of western masculinity. He began to claim permanent "residence" on his Dakota Territory ranch and sent self-serving press releases back to the New York newspapers. A short five years after his sojourn in the Dakotas, he was running for mayor of New York as a rough, tough, masculine "cowboy of the Dakotas."[30]

While Smith did not go to the extremes Roosevelt did in creating a specific masculine image, the same forces that caused Roosevelt to resort to such measures surely influenced Smith's image of his residence, his life, and himself. The most vividly expressed and forceful episodes found in Smith's recollections included the relatively short period of time he spent as a member of the Arizona Guards. Although he spent most of his time in the Guards as a miner, his memoirs of the period were dominated by his accounts of battles with Native Americans. One could attribute such a development to the fact that Indian battles and struggles to "settle" the West were a primary focus of pulp fiction and settler's firsthand accounts during the early twentieth century. It was no doubt a large part of Smith's purpose. But, daring tales of Indian battles and raids also played a large role in the formation of a western masculinity. According to David Anthony Tyeeme Clark and Joane Nagel, Native Americans were most often portrayed in contemporary accounts as a "primitive but formidable adversary" who stood in the way of western expansion and white American hegemony. Such a representation enabled American men to have a grudging respect for their foes but also allowed those who defeated the "stalwart warriors" to occupy a place as the bravest and manliest people on the continent. Thus, Native Americans were the mirror image of American hegemonic masculinity and served as the mechanism through which American men could assert their manhood.[31]

In almost all of his accounts of his fights with Native Americans, Smith described them as worthy foes but, in the end, ones that were always eventually defeated. The language he used to describe Native Americans portrayed a very masculine culture that was often oblivious to pain and was courageous in the face of death. But he also called Native Americans "savages" and an opponent that had to be dealt with harshly. In his description of the beating and whipping of Mangas Coloradas in Pinos Altos, he saved his most vivid words to create an image of a proud man who, despite torture, failed to give his enemies the satisfaction of hearing him cry out in pain. But, despite any feelings of admiration and friendship he may have had for Mangas Coloradas's bravery and masculinity, the Apache leader was still an enemy who Smith and his fellow Guards defeated and subjugated, imposing their will on a vanquished opponent. Smith may very well have viewed his years as an "Indian fighter" as the period of his life that shaped his perceptions of his manhood.[32]

Hank Smith had another motivation in presenting a romantic portrayal of his life, a motive that was pervasive throughout the American West. The most common description of the West was one of boundless opportunity, an attempt by promoters to lure settlers to the region. These same promoters, while taking great care not to discount the adventurous image of the region, had to insist that early pioneers had "tamed" the West. At the same time, some of the West's original white residents had begun publishing their reminiscences in books and periodicals and banding together in pioneer societies to sustain their conception of frontier heritage. Like Smith, their often selective memory focused on the savage wilderness they had tamed, exaggerating the past every bit as much as promoters exaggerated the present. In Crosby County, Smith was one of the original settlers. Like his counterparts throughout the West, he promoted his home region and actively sought to entice new residents to the county. His memoirs and recollections followed the same pattern than many others did throughout the American West, portraying a once wild wilderness now transformed into the perfect locale for families, businesses, and communities. Smith no doubt envisioned his life as representative of this process and served as an example of the opportunity available, thanks to men like him, in the now docile American West.[33]

After 1900 Smith began to turn the operations of the +B over to George and Robert. By 1900, he had held title to 4,431 acres of land and ran a little over three hundred head of cattle. George ran the cattle operation and Robert the sheep and farming concerns. Such a situation often created a conflict. George was a committed cattleman and did not particularly care for running sheep. George had persuaded his father to sell all his remaining sheep in 1899 so he could begin building a solely Hereford cattle operation. Robert pressed Hank to buy more sheep, since he was left managing the farm operations and occasionally helping George. Robert probably felt that his older brother overshadowed him, and the two Smith sons often fell into disputes that Smith had to settle. Generally, he sided with George and seemed to trust his judgment in managing the ranch.[34]

George Smith, like both his father and mother, was a large man, six feet two inches tall, and he weighed well over two hundred pounds. By all accounts he was a quiet, introspective man who wrote lyrical poetry during the often lonely time he spent tending to the +B herds. He inherited his father's love for books and his mother's interest in education. Most of all George was a cattleman. Besides helping his father with the Blanco Canyon animals, he also worked as a ranch hand for his cousin Fred Horsbrough at the Spur Ranch and later worked for the Half-Circle S and TM-Bar ranches, both close neighbors of the Smiths.[35]

On February 15, 1903, George Smith married Sallie Mae Adams, a native of Georgia who had moved to Texas in 1884. For the first three years of their marriage the newlyweds lived in a lean-to adjacent to the Rock House. In 1906 they moved into the Rock House to assist George's elderly parents, but two years later George built a home for his family on top of the Caprock east of the Rock House. George also followed his father into politics when, in 1908, he was elected Crosby County sheriff. He spent the weeks in Emma, while his family remained at Mount Blanco. But George disliked the time away from his family and did not stand for reelection in 1910. Sallie Smith particularly disliked George's term as sheriff; besides the time George spent away from home, she was a gentle woman who had a supreme dislike for guns and violence.[36]

George would have happily remained solely a cattleman, but circumstances meant that he had to expand and seek other economic opportunities. Besides tending his Hereford herd, he also operated a large farm. Eventually he would, like his father, serve on the Mount Blanco school board and later in the same capacity in Crosbyton. George and Sallie raised three children, daughters Evelyn and Georgia Mae, and a son, Allan. Sallie emphasized to all of her children that they receive the most education that they could. The interest in education became a passion for George Smith and his wife Sallie. Their interest in education eventually led them to move their family into Crosbyton to take advantage of the education their children could receive at the larger school in the county seat.[37]

Robert Smith was much more interested in sheep than cattle. Although the two Smith sons' experiences as sheepherders when they were children drove George to disdain the woollybacks, Robert remained a sheepherder and developed an affinity for the business. Although Elizabeth Smith would never express a favorite among her children, she did seem to share a closer bond with her youngest son. Letters and notes from Elizabeth to Robert also suggested that she sometimes worried about Robert and his direction in life. He was not near the devotee of books and intellectual pursuits that were a large part of his other sibling's lives and did not attend more than a few months of formal education. He met Barbara Massey at a Christmas Ball in Emma in 1900 and they were married by June

1901. They immediately moved into the Rock House with Hank and Elizabeth and remained there until 1904 when they moved to a half dugout a mile north of their parent's home.[38]

Robert continued to work for his father for two years after his marriage, but because of the absence of sheep on the +B, he left the ranch and set up an independent farming operation on part of the Smith landholdings. He continued to oversee the tenant farming operations for his father, primarily because George showed no interest in that part of the business and advised his father to terminate his tenant contracts and use the land as additional grazing land. Robert and Barbara raised five children, Henry, Reuben, Viola, Frank, and Floyd, in a home that Robert built on top of the Caprock in 1908.[39]

Eventually Robert and his family moved to Floydada, where he established a real estate and insurance business. Ten years later, in 1923, after the death of Elizabeth Smith, the Hank Smith estate was divided between the children; Robert received the portion that included the Rock House. So, in that year the family moved from Floydada back to Blanco Canyon. Robert returned to farming on his portion of the family holdings, and he eventually became one of the primary organizers of the West Texas Old Settlers Reunion.[40]

Hank and Elizabeth also had three daughters, Leila, Annie, and Mary. Leila was born in Fort Griffin, just before the Smith's moved to Blanco Canyon. Along with George, she went to the Estacado School for a year and finished her education at Mount Blanco. She married John Wheeler Jr., a ranch hand on the Matador Ranch, in December 1901. They had three children, Ernest, Josephine, and J. A. Eventually the family moved to Emma on homesteaded land. Of all the Smith children, Leila was perhaps the closest to family friend Charley Hawse, and after his health began to fail in 1910 he moved in with Leila and her family in Emma. Elizabeth also spent considerable time with her eldest daughter, particularly after the family moved to Emma. She would spend the winter with Leila in Emma and return to Blanco Canyon in the spring.[41]

Ann was born at the Rock House in January 1883. By the time she reached school age, the region was much more settled, and she received her education at the Mount Blanco School. The family of her eventual husband, Charles McDermett, homesteaded land just above the Caprock a short distance from the +B, and Charlese and Ann went to school together. Charles was a frequent visitor to the Smith home as a young boy, and he and Ann eventually married in 1907. Charles farmed and ranched in various places throughout the county as well as in New Mexico. He and Ann raised four children, J. Wilson, Agnes, Jeanette, and Charles Jr.[42]

Mary Magdalene was the youngest Smith child, born at Mount Blanco in 1887. In 1906, she married a local ranch hand, Vernon Lomax, an employee of the Bar-N-Bar Ranch. While all the other Smith children spent their entire lives in the South Plains region, Mary moved with her husband to Medford, Oregon, in 1913, along with daughter Louise and sons Charles and H. C. Mary and Vernon had another child, Mildred, in Oregon, but a year later, in 1914, the couple divorced and Mary and the children moved back in with Elizabeth and then to Emma in 1916. There, in 1918, Mary met and, in 1919, married, J. W. Kirk, and the couple moved onto land she inherited just northeast of the Rock House. Mary and J. W. had two children, Billy and Leila.[43]

After moving with Hank and Elizabeth to Crosby County, Charley Hawse never left the Smiths and, for all practical purposes, was a member of the family. Although he actually worked for Smith, Hawse was more of a partner than anything else. He was the one who took the lead in tending the cattle herds and, in many ways, taught Hank Smith to be a stockman. Smith came to rely on his friend for more than business purposes; he kept the operation running when Smith was away on many of his business and political ventures, and he was instrumental in helping to raise the children. Smith writes that he met Hawse while freighting in El Paso, although other Crosby County residents recounted that Smith and Charley met in California. That is more than likely not the case, since Smith's notes and memoirs never mention Hawse until he lived in Fort Griffin, where he again met his friend.[44]

Hawse never married. Except for a small interlude, he preferred to live with the Smith family and work on the +B. Perhaps following Smith's example, Hawse also entered politics for a while when he served as the Emma City marshal in the first years of the town's existence. But after approximately a year in that job he decided to move back to Blanco Canyon and return to the work on the +B. He moved in with Leila Smith in 1910 and remained with her family until his death in 1921.⁴⁵

Hank had hoped to retire from his day-to-day duties after 1900 when he turned most operations over to George and Robert, but the family suffered two setbacks during the winter of 1904–5. George fell from his horse in November of that year and was incapacitated for a few months. At one point during the ordeal, Smith wrote his friend and in-law, Fred Horsbrough, that he "feared for George's life." Hank was forced to resume the responsibilities of running the ranch during George's convalescence. Next, a devastating series of snowstorms hit the South Plains in December and January, and Smith reported the loss of over eighty-five head of cattle, including two Hereford bulls that George had previously bought to help restock the herd. The blizzards that winter were the worse that Smith had ever seen since he had moved to Blanco Canyon. Although he was now sixty-eight years old, with George injured, he had to ride out to assess the losses. Suffering from snow blindness and seeing the devastation that the blizzard had caused, Smith reported that he was "disheartened and extremely sad" to discover the herd's frozen carcasses.⁴⁶

After George's recovery, he and his father faced the prospect of having to rebuild the herd. Again, George insisted in stocking their range with Herefords. They bought another one hundred head in the spring of 1905, since the harsh winter had not only killed part of the herd, but the calf output was also low. The lack of income put a strain on the cash flow of the +B Ranch, and in 1905 Smith was forced to do something that he had vowed he would never do again—borrow money. He took a loan from the First National Bank of Floydada in the amount of $1,500. As the loan was to mature in 180 days, he obviously expected the ranch to turn around quickly.⁴⁷

However, cattle prices remained low throughout 1905, and Smith was forced to renew the loan in October and again in April 1906. That spring, with George fully recovered, the ranch's fortunes improved and Smith was able to sell seventy-five two-year-old and three-year-old steers for $25 a head and eighteen yearling steers at $18 a head, and he paid off his loan. In a letter to Horsbrough, Smith expressed his belief that the cattle market would continue strong into 1907 and that George's ideas about raising Herefords seemed sound. To take advantage of the market, Smith decided to now borrow $2,000 from the Floydada bank and purchase another one hundred head of yearlings. He continued to buy and sell cattle until his death, when his estate reported his cattle holdings as 439 head.[48]

After 1900, Smith continued to diversify his stock and did not limit his profit seeking exclusively to cattle. In 1906, he purchased an additional 456 acres adjacent to his ranch and began a hog raising operation. Smith had previously raised hogs for personal consumption, but with a good market for hogs in Amarillo, he saw an opportunity for additional profit. In a note to George, he explained that with the rise in population in the area and favorable freight rates, hogs were now a marketable enterprise. He purchased seventy-five hogs from a firm in Eastland, penned the hogs, and began his new business.[49]

Apparently, George did not share his father's enthusiasm for hog raising and strongly advised against it. Perhaps George had other reasons for trying to dissuade his father. He wrote a note to banker James Posey that offered, "thankfully the hog pens are usually upwind and I can't smell them."[50]

But Hank Smith was not a man who was swayed easily, and he took personal control over the hog operation. Despite George's misgivings, hogs turned into a profitable business for Smith. He reported a profit of $875 in 1907 and $1,052 in 1908. He did take a small loss of $55 in 1909, but in 1910, after Hank's stroke, George promptly sold all the hogs, ending the ranch's hog business.[51]

While expanding his business interests remained a primary concern for him in his later years, Smith also continued to be active in politics. He served as a Mount Blanco school trustee and actively backed county candidates that represented his visions and ideas for the region. He particularly opposed any efforts by the large cattle operations to elect their men to county posts and, given his status as the elder statesman in the region, was successful. He supported George's candidacy for Crosby County sheriff in 1908, and many observers attributed George's subsequent election to the post to his family name.[52]

Smith also took an interest in a controversial statewide issue during this time. Texas was experiencing a lively debate on the topic of prohibition. Prohibition was hardly new in the state and had been a reform issue in state politics at least since the end of the Civil War. In the 1880s, the Women's Christian Temperance Movement rallied to have the local-option system abandoned and amend the state's constitution to ban all sale of alcohol within Texas. Although the prohibitionists were not successful during the late nineteenth century, with the emergence of Progressivism in the early 1900s, particularly under the administration of Governor Thomas Campbell, prohibition emerged again as a central issue.

Prohibitionists, from their defeat in 1887 until the Campbell administration, concentrated on local-option elections with some success. This alarmed the liquor interests in the state, and in 1901 they formed the Texas Brewers Association to battle the prohibition efforts. The prohibition forces responded with the formation of the Texas Local Option Association in 1903 and a state chapter of the Anti-Saloon League in 1907. These groups merged in 1908 and began to demand that the legislature call a referendum on a statewide prohibition amendment.[53]

Smith, although he had formerly operated a bar in his hotel in Fort Griffin, was not a drinker. He decried the abuses of the saloons in Emma and did not sell alcohol in his Mount Blanco store. But, due to his German heritage and perhaps his libertine politics, he did not believe in prohibition. He joined the Anti-Statewide Prohibition Organization of Texas, and in a letter to the organization's chairman he stated that he believed the prohibition efforts "violated the principle of home rule and self-government." Such a stance put him at odds with the majority of his South Plains neighbors, but he continued to oppose prohibition until his death.[54]

He may have had another reason to oppose prohibition. Generally, during the prohibition fight, Catholics opposed any efforts to impose prohibition on alcohol within Texas. One of the facets of prohibition was anti-Catholicism among the state's predominant Protestants. Smith was born and raised a Catholic in his native Germany, but given the state's often virulent strain of anti-Catholicism, he had never publicly professed his religion since coming to Texas.

So, Smith's opposition to prohibition may also have been due to his familial Catholicism.[55]

Hank Smith suffered a light stroke in February 1910. It curtailed some of his activity but it was not considered debilitating. He did suffer from some lack of mobility, and the condition prevented him from taking an active interest in the ranch and kept him from finishing his memoirs. Because Charlie Hawse was also in failing health, George was forced to hire new hands to help him keep the ranch operational. Smith convalesced throughout the winter, often sitting for hours on the porch of the Rock House and staring at the canyon that surrounded his home. He was no doubt contemplative; he had expected to finish and publish his memoirs, but the stoke had left him able to write only short sentences. His thoughts drifted to the Blanco Canyon and South Plains region; it had been almost devoid of people when he arrived but now boasted towns, farms, businesses, and ranches. Smith probably did realize his importance in the region and how far he had traveled from his boyhood German home.[56]

Smith, now seventy-four years old, realized that his time was growing short. Although he had traveled great distances in his life, it had always been with a purpose, to start a new business or to escape another one. What he had never been able to do was take a trip for the sheer joy of it, and he had not been back to visit his sister since he left Ohio at the age of sixteen. So in the summer of 1910, at Elizabeth's urging, the Smiths left Texas for an extended trip to visit relatives in Ohio and Massachusetts.[57]

Fred Horsbrough took the Smiths by wagon to Amarillo where they boarded a train for Sandusky, Ohio, where Smith's sister Mary had settled. Smith's two older sisters, Ann and Margaret, had already died and he had not seen Mary in almost sixty years. His children had never met his closest relatives, and he expressed a "feeling of great sadness" that they could not accompany Elizabeth and him on this trip. Mary was overjoyed to see her brother when they arrived, but amazingly they found that they needed an interpreter to speak. Hank, who had learned to speak English, Chiricahua, and Spanish, had lost the ability to speak German. Mary spoke very little English and her husband none at all. Nevertheless, they carried on as best they could, and Hank called his visit with his sister the "completion of his life." The Smiths also visited Cleveland and were awestruck at the sight this large city. Elizabeth's postcards home expressed wonderment that so many people could live in one place. The Smiths were particularly taken with their visit to a swimming pool in Cleveland. They were very surprised that people would gather (and pay) simply to enjoy the water. Smith thought the whole idea was a huge waste and remarked that he could remember many times when he would have "loved to have just a part" of the pool to water his "thirsty stock."[58]

The Smiths next boarded another train and traveled to Cochituate, Massachusetts, where Elizabeth had relatives. They spent a month at the place, so they must have enjoyed themselves. From Massachusetts, the Smiths returned to Texas, but because Hank was so exhausted from the trip the couple spent the remainder of the summer in Amarillo with Fred Horsbrough. Hank was now seventy-five, but a story in the *Amarillo News* reported him "as spry and far more stable than many men of forty. He is in robust health with a promise of still greater age."[59]

During the trip north, Smith never forgot that his business fortunes were still in Texas. While Elizabeth's letters home (usually addressed to Robert) were full of details, such as the fortunes of her family and the places they visited, Hank's letters contained more sober matters. He addressed his letters to George, who was continuing to manage the ranch. Hank did not bother with the details of the trip, instead he inquired about the cattle, prices, how the finances of the operation were, and assured George that he could "come home as quick as possible if he was needed." He asked about the weather and if the stock had enough water. He lectured George not to "get taken on any prices" and showed particular concern that George was keeping up with his meteorological records.[60]

Despite the *Amarillo News*'s glowing report on Smith's health, the combination of the stroke and the long trip took a physical toll on Hank. He had always enjoyed a keen interest in buying and selling cattle and horses, even though George was managing the ranch operation. But when Horsbrough wrote him in June 1911, informing him of a potential buyer for his cattle, Smith replied that the buyer would have to "discuss it with George" since he now very rarely left the house. He gave up his position as Mount Blanco school trustee and even stopped traveling to the new county seat of Crosbyton. Elizabeth noted that he liked "to just sit in his chair on the porch" and hardly ever moved. He even quit tending to his beloved fruit trees that lined the road that led to his home, something that he had always taken great pride in. The winter was unduly harsh on him, and he found that he could barely rise from his bed. He remarked to a visitor that he would "surely not see another winter." Finally, on May 19, 1912, Hank Smith died, apparently of a heart attack.[61]

Hank Smith's May 25, 1912, obituary in the *Crosbyton Review* noted that "with his passing surely a chapter has closed in the South Plains and the country has lost a truly magnificent man." He left an estate that was estimated to be worth anywhere between five hundred thousand and a million dollars, most of it in land. Elizabeth continued to live in the Rock House in the next few years after Hank's death, but she was finally forced to give up her job as Mount Blanco postmistress in 1916. After that she spent most of her time living with Leila and her family, although she always returned to Blanco Canyon for the summer. Charlie Hawse also moved in with Leila until his death in 1921. Elizabeth passed away in June 1925.[62]

Many years after his death, Smith's granddaughter Georgia Mae Ericson, George's youngest child, discovered that Hank Smith was born and baptized a Catholic. Ericson was concerned that because he had not revealed that fact after moving to Texas, he had died without receiving the last rites. She contacted Father Malcolm Neyland, who was instrumental in establishing Catholic churches in Ralls and Crosbyton and inquired whether there was some way to give the last rites after a person's death. Father Neyland assured her that it was within Church canons and he would be glad to perform such a duty for her grandfather. So, on Smith's 150th birthday, in the Emma cemetery, Hank Smith was given the last rites, almost seventy-five years after his death.[63]

During his life and continuing after his death, Henry C. "Hank" Smith became a legendary figure of the South Plains. Newspapers, scholarly articles, and Hubert Curry's biography exalted him as emblematic of the romance and adventure of the American West.

Certainly, Hank Smith was an important figure in the settling of the area, and he occupies an important place. However, more than anything else Smith represented a more vital place in American western history, one that did much more to shape the region than the romantic tales that often dominate the nation's collective memory. He was a practical businessman who came to the West to seek economic opportunity. Enterprise was his main focus in life, a focus that never changed. He lived an adventurous life, but he was not an adventurer. Rather, he was an entrepreneur, something that is often adventurous enough. His life is an example of what was the most typical of a westering man, a man who journeyed the West not for the romance of the open country, but for the sometimes-cold reality of profit.

Epilogue

Many of the places that Hank Smith passed through, or lived in, no longer exist, or are mere shadows of what they formerly were. Pinos Altos, New Mexico, is a veritable ghost town with little to suggest the flurry of activity that once swirled around the mining community. Forts Quitman and Griffin are nothing more than ruins, and the bustling town around Fort Griffin, The Flat, that gave Smith his first measure of economic success, has completely disappeared. The community of Mount Blanco and the once boomtown of Emma are also completely gone. Crosby County's first county seat, Estacado, now contains no more than a few abandoned buildings, a dilapidated cotton gin, some farmhouses, and a Texas State Historical Marker that only hints at the town's former prominence. The new county seat, Crosbyton, once a prosperous agricultural community, now struggles, like so many rural Texas towns, to maintain population and a viable economy. Even Hank Smith's Rock House, once a refuge for many travelers making their way across the desolate South Plains and a majestic beacon, now sits in ruins from a 1950s fire and on private land away from any public access.

Blanco Canyon, an area that Elizabeth Smith once termed a "garden spot," is now almost completely devoid of water, the victim of human interaction, careless irrigation, and a changing climate. The white cliffs of Mount Blanco have now been leveled off and are almost unrecognizable as a unique feature on the landscape, with no hint of a legacy as a landmark. Hank's fruit trees no longer remain and only a few of the elegant bois d'arc's stand guard over the road to the Rock House. In some ways, Blanco Canyon has reverted to the virtually uninhabited state that greeted the Smiths when they moved there in 1878. But its reversion is not back to a natural state, and it bears the scars of over a hundred years of cultivation, grazing, and overproduction.

Ranches do remain in the Blanco Canyon region, but they, like the towns and land, are minuscule compared to their former greatness. The McNeil SR Ranch remains, but at a greatly reduced amount of acreage. The once-mighty Matador still operates, but on onetenth the grazing land and in a completely new set of market conditions. If he could see his former home today, Hank Smith would recognize it, but it would be a poor reflection.

Crosby County has ably attempted to maintain its heritage and recognize the contributions of Hank Smith in its legacy. In the county's small but excellent Pioneer Memorial Museum, there is an entire room devoted to the life of Hank Smith and his importance to the region; indeed, the museum facility itself is a replica of Hank's Rock House. As one enters the Hank Smith room, one first sees a large mural depicting Smith's life. Written in bold script at the top of the mural is "Heinrich Schmitt was born in Germany and immigrated to America in 1851 to become Uncle Hank Smith and live the adventurous life of the Old West." The various vignettes in the wall painting illustrate the numerous events in Smith's life, such as his time spent as a "bullwhacker" on the Santa Fe Trail, "busting broncs" in California, engaging in duels, panning for gold, fighting Indians, and meeting notorious western characters, such as Billy the Kid and Judge Roy Bean. The mural is designed to give the impression that this was a man who lived the life of the "Old West" as it was portrayed in film, television, western novels, and our nation's memory.[1]

But as one peruses the exhibits on Smith's life, a different impression comes into focus. There are family photographs depicting the life of a devoted family man, exhibits describing his various business ventures, his experiences in helping organize Crosby County, and, to some, mere mundane activities. Clearly they reveal that Hank Smith's life was much more than a series of adventures. It was rather one of an ordinary man who lived an ordinary life that occasionally included some extraordinary encounters.

Smith's life, his struggles to establish a viable business, and his eventual success with the Cross B Ranch is a testament to his fortitude and, maybe more, his obstinate personality. Hank Smith was firmly committed to securing economic security, and his dogged pursuit of such a goal shaped his life. But even in eventually accomplishing that security, he was also ordinary. Although he may have achieved more success than many who settled in the West, and not as much as a fewer number of others, his pursuit of profits was typical of the motivations that drew the majority of people to the West. Perhaps it would be most appropriate to say that Henry C. "Hank" Smith was the quintessential westering man.

Notes

Chapter 1

1. Richard White, *"It's Your Misfortune and None of My Own": A New History of the American West* (Norman: University of Oklahoma Press, 1991), 285; Patricia Nelson Limerick, *A Legacy of Conquest: The Unbroken Past of the American West* (New York: W. W. Norton, 1987).

Chapter 2

1. *Crosbyton Review*, May 23, 1912, 1, 2.
2. David Blackbourn, *The Long Nineteenth Century: A History of Germany, 1780–1918* (Oxford: Oxford University Press, 1998), 13, 28, 93, 96–98.
3. Blackbourn, *Long Nineteenth Century*, 21, 27, 141, 161–62, 239.
4. Blackbourn, *Long Nineteenth Century*, 110–12.
5. The Henry C. "Hank" Smith Papers, Folders 1 and 2, Historic Research Center, Panhandle-Plains Historical Museum, Canyon, TX; Blackbourn, *Long Nineteenth Century*, 112–13.
6. Henry C. "Hank" Smith Papers, Folder 2; W. Hubert Curry, *Sun Rising on the West: The Saga of Henry Clay and Elizabeth Smith* (Crosbyton, TX: Crosby County Pioneer Memorial, 1979), 9–11.
7. Curry, *Sun Rising on the West*, 9–11.
8. Henry C. "Hank" Smith Papers, Folder 2.
9. Blackbourn, *Long Nineteenth Century*, 111–14.
10. Henry C. "Hank" Smith Papers, Folder 2; Blackbourn, *Long Nineteenth Century*, 106–7.
11. Henry C. "Hank" Smith Papers, Folder 2.
12. Henry C. "Hank" Smith Papers, Folder 2.
13. Blackbourn, *Long Nineteenth Century*, 191–92.
14. Blackbourn, *Long Nineteenth Century*, 194.
15. Henry C. "Hank" Smith Papers, Folder 3; Don Heinrich Tolzmann, *The German-American Experience* (New York: Humanity Books, 2000), 151–52, 159.
16. Henry C. "Hank" Smith Papers, Folder 3.
17. Blackbourn, *Long Nineteenth Century*, 196–97; Tolzmann, *German-American Experience*, 160–61.
18. Tolzmann, *German-American Experience*, 193–97; Larry L. Miller, *Ohio Place Names* (Bloomington: Indiana University Press, 1996), 190; Henry C. "Hank" Smith Papers, Folder 4.

19. George W. Knepper, *Ohio and Its People* (Kent, OH: Kent State University Press, 1989), 122–26; Francis P. Weisenburger, *The History of the State of Ohio*, vol. 3, *The Passing of the Frontier, 1825–1850* (Columbus: Ohio State Archaeological and Historical Society, 1941), 4–8; Henry C. "Hank" Smith Papers, Folder 5.
20. Henry C. "Hank" Smith Papers, Folder 5.
21. Weisenburger, *History of the State of Ohio*, 3: 104–6, 197.
22. Henry C. "Hank" Smith Papers, Folder 5.
23. Curry, *Sun Rising on the West*, 12.
24. David Dary, *The Santa Fe Trail: Its History, Legends, and Lore* (New York: Penguin Books, 2002), 239.
25. Susan Calafate Boyle, *Los Capitalistas: Hispano Merchants and the Santa Fe Trade* (Albuquerque: University of New Mexico Press, 1997), 90–93.
26. Henry Pickering Walker, *The Wagonmasters: High Plains Freighting from the Earliest Days of the Santa Fe Trail to 1880* (Norman: University of Oklahoma Press, 1966), 77–78.
27. Walker, *Wagonmasters*, 77–78.
28. Walker, *Wagonmasters*, 23–24; Dary, *Santa Fe Trail*, 220.
29. Walker, *Wagonmasters*, 23–24.
30. Walker, *Wagonmasters*, 84–85.
31. Walker, *Wagonmasters*, 29; Henry C. "Hank" Smith Papers, Folder 5.
32. Henry C. "Hank" Smith Papers, Folder 6; Curry, *Sun Rising on the West*, 14.
33. Henry C. "Hank" Smith Papers, Folder 6.
34. Henry C. "Hank" Smith Papers, Folder 6; Walker, *Wagonmasters*, 46.
35. Walker, *Wagonmasters*, 46–47.
36. Henry C. "Hank" Smith Papers, Folder 7.
37. Raymond W. Settle and Mary L. Settle, *War Drums and Wagon Wheels: The Story of Russell, Majors, and Waddell* (Lincoln: University of Nebraska Press, 1966), 235–56.
38. Henry C. "Hank" Smith Papers, Folder 7; Leonard J. Arrington and Davis Bitton, *The Mormon Experience: A History of the Latter-Day Saints* (New York: Knopf, 1979), 87–94.
39. Henry C. "Hank" Smith Papers, Folder 7.
40. Henry C. "Hank" Smith Papers, Folder 7.
41. Henry C. "Hank" Smith Papers, Folder 7.
42. Henry C. "Hank" Smith Papers, Folder 7.
43. Henry C. "Hank" Smith Papers, Folder 7.
44. Henry C. "Hank" Smith Papers, Folder 7.

Chapter 3

1. Henry C. "Hank" Smith Papers, Folder 38; Hattie M. Anderson, "Mining and Indian Fighting in Arizona and New Mexico, 1858–1861: Memoirs of Hank Smith," *Panhandle-Plains Historical Review* 1 (1928): 74.
2. Henry C. "Hank" Smith Papers, Folder 38; Rodman W. Paul, *The Far West and the Great Plains in Transition: 1859–1900* (New York: Harper and Row, 1988), 55–56.

3. Paul, *Far West and the Great Plains in Transition*, 56–57; Roscoe P. Conkling and Margaret B. Conkling, *The Butterfield Overland Mail: 1857–1869* (Glendale, CA: A. H. Clark, 1947), 6–12.

4. Paul, *Far West and the Great Plains in Transition*, 56–57.

5. Henry C. "Hank" Smith Papers, Folder 38; Jay J. Wagoner, *Early Arizona* (Tucson: University of Arizona Press, 1975), 354.

6. Henry C. "Hank" Smith Papers, Folder 38; Richard J. Perry, *Western Apache Heritage: People of the Mountain Corridor* (Austin: University of Texas Press, 1991), 146–47, 170.

7. Perry, *Western Apache Heritage*, 170–71.

8. Perry, *Western Apache Heritage*, 171; Marshall Trimble, *Arizona: A Panoramic History of a Frontier State* (New York: Doubleday, 1977), 154.

9. Henry C. "Hank" Smith Papers, Folder 38. Wagoner, *Early Arizona*, 355–56.

10. Henry C. "Hank" Smith Papers, Folder 39.

11. David Dary, *Entrepreneurs of the Old West* (New York: Alfred A. Knopf, 1986), 324–25.

12. Henry C. "Hank" Smith Papers, Folder 39.

13. Edwin R. Sweeney, *Mangas Coloradas: Chief of the Chiricahua Apaches* (Norman: University of Oklahoma Press, 1998), xv.

14. Henry C. "Hank" Smith Papers, Folder 41; Sweeney, *Mangas Coloradas*, 384–85.

15. Henry C. "Hank" Smith Papers, Folder 44; Sweeney, *Mangas Coloradas*, 385–87; Patricia F. Meleski, *Echoes of the Past—New Mexico's Ghost Towns* (Albuquerque: University of New Mexico Press, 1972), 189–96; Fayette Alexander Jones, *New Mexico Mines and Minerals* (Santa Fe: New Mexico Printing, 1904). Both Meleski and Jones confirm that it was Snively, Birch, and Hicks who made the 1860 discovery of gold in Pinos Altos, and neither one mentions Smith's presence.

16. Henry C. "Hank" Smith Papers, Folder 44; Curry, *Sun Rising on the West*, 41–45.

17. Curry, *Sun Rising on the West*, 40–41; Meleski, *Echoes of the Past—New Mexico's Ghost Towns*, 190–94; James Tevis, "Arizona in the 50s," *True West Magazine*, June–August 1968, 10–11.

18. Sweeney, *Mangas Coloradas*, 387; Tevis, "Arizona in the 50s."

19. Sweeney, *Mangas Coloradas*, 391.

20. Henry C. "Hank" Smith Papers, Folder 46; Sweeney, *Mangas Coloradas*, 399–400.

21. Henry C. "Hank" Smith Papers, Folder 46.

22. Henry C. "Hank" Smith Papers, Folder 46.

23. Henry C. "Hank" Smith Papers, Folder 46; Sweeney, *Mangas Coloradas*, 400–403; John C. Cremony, *Life among the Apaches* (Lincoln: University of Nebraska Press, 1983), 173–74.

24. Sweeney, *Mangas Coloradas*, 417; Edwin R. Sweeney, *Cochise: Chiricahua Apache Chief* (Norman: University of Oklahoma Press, 1991), 189–90.

25. Sweeney, *Mangas Coloradas*, 417.

26. Henry C. "Hank" Smith Papers, Folder 48; Curry, *Sun Rising on the West*, 50–51.

27. Sweeney, *Mangas Coloradas*, 416–17.
28. Sweeney, *Mangas Coloradas*, 417.
29. Henry C. "Hank" Smith Papers, Folder 50; Sweeney, *Mangas Coloradas*, 415, 417.
30. Sweeney, *Mangas Coloradas*, 419–20.
31. Henry C. "Hank" Smith Papers, Folder 50; Sweeney, *Mangas Coloradas*, 420; George Wythe Baylor, *John Robert Baylor: Confederate Governor of Arizona* (Tucson: Arizona Pioneers' Historical Society, 1966), 12.
32. Sweeney, *Mangas Coloradas*, 420.
33. Sweeney, *Mangas Coloradas*, 420–21; Henry C. "Hank" Smith Papers, Folder 52.
34. Sweeney, *Mangas Coloradas*, 420–21.
35. Sweeney, *Mangas Coloradas*, 422; Robert Lee Kerby, *The Confederate Invasion of New Mexico and Arizona, 1861–1862* (Tucson: Westernlore Press, 1981), 29–36; Donald S. Frazier, *Blood and Treasure: Confederate Empire in the Southwest* (College Station: Texas A&M University Press, 1995), 50–52.
36. Sweeney, *Mangas Coloradas*, 423–24.
37. Sweeney, *Mangas Coloradas*, 424; Henry C. "Hank" Smith Papers, Folder 53.
38. Henry C. "Hank" Smith Papers, Folder 53.
39. Henry C. "Hank" Smith Papers, Folder 53.
40. Henry C. "Hank" Smith Papers, Folder 54.

Chapter 4

1. Jerry Thompson, *Civil War in the Southwest: Recollections of the Sibley Brigade* (College Station: Texas A&M University Press, 2001), 3.
2. Donald S. Frazier, *Blood and Treasure: Confederate Empire in the Southwest* (College Station: Texas A&M University Press, 1995), 18–19. The declared territory of Arizona did not share the same borders of the present-day state of Arizona. The delegates who had met in Tucson to declare the new territory drew the boundaries to incorporate the southern half of the present-day states of Arizona and New Mexico, a boundary that bordered Mexico on the south and stretched east to west from Texas to California.
3. Frazier, *Blood and Treasure*, 19–20.
4. Frazier, *Blood and Treasure*, 21.
5. Jerry Thompson, *Colonel John Robert Baylor: Texas Indian Fighter and Confederate Soldier* (Hillsboro, TX: Hill Junior College Press, 1971), 24–31.
6. Thompson, *Colonel John Robert Baylor*, 4–7.
7. Thompson, *Colonel John Robert Baylor*, 4–7.
8. Thompson, *Colonel John Robert Baylor*, 4–7.
9. Frazier, *Blood and Treasure*, 27.
10. William Davidson, "Organization of the Sibley Brigade," in *Civil War in the Southwest: Recollections of the Sibley Brigade*, ed. Jerry Thompson (College Station: Texas A&M University Press, 2001), 4–5.
11. Frazier, *Blood and Treasure*, 41–43.

12. Henry C. "Hank" Smith Papers, Folder 61; Curry, *Sun Rising on the West*, 71; Frazier, *Blood and Treasure*, 31–33.
13. Frazier, *Blood and Treasure*, 48.
14. Frazier, *Blood and Treasure*, 54–57.
15. Henry C. "Hank" Smith Papers, Folder 61.
16. Henry C. "Hank" Smith Papers, Folder 61; Frazier, *Blood and Treasure*, 57.
17. Frazier, *Blood and Treasure*, 57; Curry, *Sun Rising on the West*, 72.
18. Henry C. "Hank" Smith Papers, Folder 62; Frazier, *Blood and Treasure*, 57–58.
19. Henry C. "Hank" Smith Papers, Folder 62.
20. Henry C. "Hank" Smith Papers, Folder 63.
21. Henry C. "Hank" Smith Papers, Folder 63.
22. Henry C. "Hank" Smith Papers, Folder 63; Frazier, *Blood and Treasure*, 59.
23. Henry C. "Hank" Smith Papers, Folder 63.
24. Henry C. "Hank" Smith Papers, Folder 63; Frazier, *Blood and Treasure*, 59–60.
25. Frazier, *Blood and Treasure*, 60–61.
26. Frazier, *Blood and Treasure*, 60–61.
27. Henry C. "Hank" Smith Papers, Folder 64.
28. Frazier, *Blood and Treasure*, 110.
29. Frazier, 111.
30. Frazier, 111.
31. Frazier, 112.
32. Frazier, 113.
33. Frazier, 113.
34. Henry C. "Hank" Smith Papers, Folder 65; Frazier, *Blood and Treasure*, 113.
35. Henry C. "Hank" Smith Papers, Folder 65.
36. Henry C. "Hank" Smith Papers, Folder 65; Frazier, *Blood and Treasure*, 114.
37. Frazier, *Blood and Treasure*, 114–16.
38. Henry C. "Hank" Smith Papers, Folder 66; Sweeney, *Mangas Coloradas*, 426–27.
39. Sweeney, *Mangas Coloradas*, 427; Frazier, *Blood and Treasure*, 190.
40. Henry C. "Hank" Smith Papers, Folder 67; Frazier, *Blood and Treasure*, 190.
41. Henry C. "Hank" Smith Papers, Folder 67. Sweeney, *Mangas Coloradas*, 190.
42. Sweeney, *Mangas Coloradas*, 190.
4.3 Frazier, *Blood and Treasure*, 191.
44. Frazier, *Blood and Treasure*, 196–97.
45. Henry C. "Hank" Smith Papers, Folder 68.
46. Frazier, *Blood and Treasure*, 208–58.
47. Curry, *Sun Rising on the West*, 102–3.
48. Henry C. "Hank" Smith Papers, Folder 70; Curry, *Sun Rising on the West*, 103.
49. Henry C. "Hank" Smith Papers, Folder 71.
50. Henry C. "Hank" Smith Papers, Folder 70; Curry, *Sun Rising on the West*, 105–6.

Chapter 5

1. Henry C. "Hank" Smith Papers, Folder 83.
2. Dary, *Santa Fe Trail*, 281–91.
3. Henry C. "Hank" Smith Papers, Folder 83.
4. Dary, *Santa Fe Trail*, 281–91.
5. W. H. Timmons, *El Paso: A Borderlands History* (El Paso: University of Texas at El Paso, 1990), 135–39.
6. Timmons, *El Paso*, 145–46.
7. Earl W. Heathcote, "Business of El Paso," in *El Paso: A Centennial Portrait*, ed. Harriot Howze Jones (El Paso: El Paso Historical Society, 1972), 192–96.; Timmons, *El Paso*, 160–61.
8. Gama Loy Christian, "Sword and Plowshare: The Symbiotic Development of Port Bliss and El Paso, Texas, 1849–1918 (PhD diss., Texas Tech University, 1977), 30–31.
9. Timmons, *El Paso*, 162.
10. Henry C. "Hank" Smith Papers, Folder 88.
11. Henry C. "Hank" Smith Papers, Folder 88.
12. Timmons, *El Paso*, 159–60.
13. Timmons, *El Paso*, 143, 155; Curry, *Sun Rising on the West*, 114.
14. Henry C. "Hank" Smith Papers, Folder 89.
15. Henry C. "Hank" Smith Papers, Folder 89.
16. Henry C. "Hank" Smith Papers, Folder 89.
17. Henry C. "Hank" Smith Papers, Folder 89.
18. Henry C. "Hank" Smith Papers, Folder 90.
19. Henry C. "Hank" Smith Papers, Folder 91.
20. Ty Cashion, *A Texas Frontier: The Clear Fork Country and Fort Griffin, 1849–1887* (Norman: University of Oklahoma Press, 1996), 81–99, 134–36.
21. John Stricklin Spratt, *The Road to Spindletop: Economic Change in Texas, 1875–1901* (Dallas: Southern Methodist University Press, 1955), 37–48.
22. Cashion, *A Texas Frontier*, xii–xiv.
23. Cashion, 174.
24. Cashion, 174–75.
25. Cashion, 174–75.
26. Cashion, 160.
27. Henry C. "Hank" Smith Papers, Folder 95. Smith reports that the previous contract holder was a W. H. Hicks. This is probably a mistake and it was more than likely W. B. Hicks, a former post sutler and trader in the area.
28. Henry C. "Hank" Smith Papers, Folder 95.
29. Henry C. "Hank" Smith Papers, Folder 95.
30. Henry C. "Hank" Smith Papers, Folder 95.
31. John Hutto, "Mrs. Elizabeth (Aunt Hank) Smith," *West Texas Historical Association Yearbook*, vol. 15 (1939), Abilene, TX, 40–42.
32. Hutto, "Mrs. Elizabeth (Aunt Hank) Smith"; Henry C. "Hank" Smith Papers, Folder 94; Curry, *Sun Rising on the West*, 129.
33. Henry C. "Hank" Smith Papers, Folder 94; Cashion, *A Texas Frontier*, 161–62.

34. Cashion, *A Texas Frontier*, 162.
35. Cashion, 162–63.
36. Cashion, 162–63.
37. Henry C. "Hank" Smith Papers, Folder 96.
38. Henry C. "Hank" Smith Papers, Folder 96; Curry, *Sun Rising on the West*, 131.
39. Henry C. "Hank" Smith Papers, Folder 96.
40. Henry C. "Hank" Smith Papers, Folder 96; Cashion, *A Texas Frontier*, 176.
41. Henry C. "Hank" Smith Papers, Folder 97; Curry, *Sun Rising on the West*, 142.
42. *Crosbyton (TX) Review*, February 29, 1912, 1, 4, 6.
43. Henry C. "Hank" Smith Papers, Folder 98; Curry, *Sun Rising on the West*, 145. Biggers's recollection of Blanco River is more than likely what became White River.
44. Henry C. "Hank" Smith Papers, Folder 97.
45. Henry C. "Hank" Smith Papers, Folder 97.
46. Henry C. "Hank" Smith Papers, Folder 97; Curry, *Sun Rising on the West*, 149.
47. Henry C. "Hank" Smith Papers, Folder 97.
48. Henry C. "Hank" Smith Papers, Folder 97; Curry, *Sun Rising on the West*, 146.
49. Henry C. "Hank" Smith Papers, Folder 97.
50. Henry C. "Hank" Smith Papers, Folder 97.

Chapter 6

1. "Blanco Canyon," *Handbook of Texas Online*, the Texas State Historical Association, 1997–2002. Some early maps misidentified the White River as the Salt Fork of the Brazos, which flows into the White River below Blanco Canyon.
2. "Blanco Canyon," *Handbook of Texas Online*; Ernest Wallace, *Ranald S. Mackenzie on the Texas Frontier* (Lubbock: West Texas Museum Association, 1964), 45–57.
3. Wallace, *Ranald S. MacKenzie on the Texas Frontier*, 58–63.
4. Wallace, 58–63.
5. Wallace, 58–63.
6. Henry C. "Hank" Smith Papers, Folder 100; Curry, *Sun Rising on the West*, 149–50, 153.
7. Henry C. "Hank" Smith Papers, Folder 101.
8. Henry C. "Hank" Smith Papers, Folder 101.
9. Henry C. "Hank" Smith Papers, Folder 101; Hutto, "Mrs. Elizabeth (Aunt Hank) Smith," 45–47.
10. Henry C. "Hank" Smith Papers, Folder 102.
11. Henry C. "Hank" Smith Papers, Folder 101; Curry, *Sun Rising on the West*, 161.
12. Dan Flores, *Caprock Canyonlands: Journeys into the Heart of the Southern Plains* (Austin: University of Texas Press, 1990), 55–56.
13. Henry C. "Hank" Smith Papers, Folder 102.

14. Curry, *Sun Rising on the West*, 156–57; Hutto, "Elizabeth (Aunt Hank) Smith," 47.

15. Curry, *Sun Rising on the West*, 261.

16. Henry C. "Hank" Smith Papers, Folder 102; W. M. Pearce, *The Matador Land and Cattle Company* (Norman: University of Oklahoma Press, 1964), 23–34. Spottswood Lomax, who admired Spanish literature, is credited with naming the new company the Matador.

17. Pearce, *Matador Land and Cattle Company*, 59–128.

18. Pearce, *Matador Land and Cattle Company*, 59–128; Henry C. "Hank" Smith Papers, Folder 103; Hubert Curry, *Sun Rising on the West*, 261–62.

19. William Curry Holden, *The Espuela Land and Cattle Company: A Study of a Foreign-Owned Ranch in Texas* (Austin: Texas State Historical Association, 1970), 23–141.

20. Holden, *Espuela Land and Cattle Company*, 23–141.

21. *A History of Crosby County, 1876–1977* (Crosbyton, TX: Crosby County Historical Commission, 1978), 386; Curry, *Sun Rising on the West*, 262.

22. Curry, *Sun Rising on the West*, 264.

23. Curry, *Sun Rising on the West*, 264–65.

24. J. C. McNeill III, *The McNeill's SR Ranch: 100 Years in Blanco Canyon* (College Station: Texas A&M University Press, 1988), 47–58.

25. McNeill, *McNeill's SR Ranch*, 47–58.

26. Curry, *Sun Rising on the West*, 265–67.

27. Henry C. "Hank" Smith Papers, Folder 104; John Cooper Jenkins, *Estacado: Cradle of Culture and Civilization on the Staked Plains of Texas* (Crosbyton, TX: Crosby County Pioneer Memorial Museum, 1986), 31–38.

28. Henry C. "Hank" Smith Papers, Folder 104.

29. Henry C. "Hank" Smith Papers, Folder 105.

30. Henry C. "Hank" Smith Papers, Folder 105.

31. Henry C. "Hank" Smith Papers, Folder 105; Curry, *Sun Rising on the West*, 168–70.

32. Terry G. Jordan, *North American Cattle-Ranching Frontiers: Origins, Diffusions, and Differentiation* (Albuquerque: University of New Mexico Press, 1993), 210–11.

33. Robert V. Hine and Edwin R. Bingham, eds., "The Cattle Frontier," in *The American Frontier: Readings and Documents* (Boston: Little, Brown, 1971), 216.

34. Hine and Bingham, "Cattle Frontier," 216.

35. Henry C. "Hank" Smith Papers, Folder 108.

36. Henry C. "Hank" Smith Papers, Folder 108; Curry, *Sun Rising on the West*, 183.

37. Edward Everett Dale, *The Range Cattle Industry: Ranching on the Great Plains from 1865 to 1925*, 3rd ed. (Norman: University of Oklahoma Press, 1960), 106–7.

38. Dale, *Range Cattle Industry*; Henry C. "Hank" Smith Papers, Folders 109 and 111; Kentucky Cattle Raising Company Records, 1886–1894, Southwest Collection, Texas Tech University.

39. Henry C. "Hank" Smith Papers, Folder 181; Crosby County Deed Record Book II, Crosby County Clerk's Office, Crosbyton, TX, 619.

40. Henry C. "Hank" Smith Papers, Folder 124; Dale, *Range Cattle Industry*, 109.

41. Henry C. "Hank" Smith Papers, Folder 131; Espuela Land and Cattle Company Ltd., Records, Southwest Collection, Texas Tech University.

42. Henry C. "Hank" Smith Papers, Folder 130.

43. Henry C. "Hank" Smith Papers, Folder 193.; Dale, *Range Cattle Industry*, 150; Jordan, *North American Cattle-Ranching Frontiers*, 270.

44. Henry C. "Hank" Smith Papers, Folder 107; Paul H. Carlson, *Texas Woollybacks: The Range Sheep and Goat Industry* (College Station: Texas A&M University Press, 1982), 66, 71.

45. Carlson, *Texas Woollybacks*, 48–49, 66.

46. Carlson, *Texas Woollybacks*, 67–69.

47. Carlson, *Texas Woollybacks*, 75–76; Henry C. "Hank" Smith Papers, Folder 168; Scott Sosebee, "Henry C. 'Hank' Smith: A Portrait of a South Plains Capitalist," *West Texas Historical Association Yearbook*, 77 (2001): 76–77.

48. Curry, *Sun Rising on the West*, 184–85.

49. Curry, *Sun Rising on the West*, 184–85; Henry C. "Hank" Smith Papers, Folder 121.

50. Carlson, *Texas Woollybacks*, xii; McNeill, *McNeil SR Ranch*, 108; Henry C. "Hank" Smith Papers, Folder 115.

51. Henry C. "Hank" Smith Papers, Folder 114; Limerick, *Legacy of Conquest*, 68.

52. Henry C. "Hank" Smith Papers, Folder 117; Curry, *Sun Rising on the West*, 186.

53. Henry C. "Hank" Smith Papers, Folder 123; Sosebee, "Henry C. 'Hank' Smith: Portrait of a South Plains Capitalist," 77–78.

54. Sosebee, "Henry C. 'Hank' Smith: A Portrait of a South Plains Capitalist," 78.

55. Sosebee, 78–79.

56. Sosebee, 79–80.

57. Sosebee, 79–80.

58. Dary, *Entrepreneurs of the Old West*, 324–25; Sosebee, "Henry C. 'Hank' Smith: A Portrait of a South Plains Capitalist," 83–84.

Chapter 7

1. Henry C. "Hank" Smith Papers, Folder 157; Curry, *Sun Rising on the West*, 197–98. Livermore and Smith's claim of a vast underground water resource is the Ogallala Aquifer, a water formation that spreads under more than 35,000 square miles of the Great Plains. However, the aquifer did not prove inexhaustible, as irrigation and a lack of recharge sources have greatly diminished its capacity. The springs that he reported no longer exist and the White River no longer flows through Blanco Canyon.

2. Henry C. "Hank" Smith Papers, Folder 160.

3. David Wrobel, *Promised Lands: Promotion, Memory, and the Creation of the American West* (Lawrence: University Press of Kansas, 2002), 2.

4. Wrobel, *Promised Lands*, 5–6.

5. Wrobel, *Promised Lands*, 5–6.

6. Jenkins, Estacado, 45–52; Curry, *Sun Rising on the West*, 192–93.

7. *History of Crosby County*, 15–16.

8. Henry C. "Hank" Smith Papers, Folder 161; *History of Crosby County*, 10.

9. Henry C. "Hank" Smith Papers, Folder 161.

10. Henry C. "Hank" Smith Papers, Folder 162.

11. Henry C. "Hank" Smith Papers, Folder 162; Curry, *Sun Rising on the West*, 204; Sarah Ann Britton, *The Early History of Baylor County* (Dallas: Story Book Press, 1955), 41.

12. Curry, *Sun Rising on the West*, 206; Crosby County Election Records, Crosby County Clerk's Office, Book 1.

13. Henry C. "Hank" Smith Papers, Folder 181; Curry, *Sun Rising on the West*, 206.

14. Dundee Division, Matador Land and Cattle Company Records, Box 23, Folder 2, Southwest Collection/Special Collections, Texas Tech University, Lubbock.

15. Dundee Division, Matador Land and Cattle Company Records, Box 23, Folder 2, Southwest Collection/Special Collections, Texas Tech University, Lubbock; Henry C. "Hank" Smith Papers, Day Ledger no. 1; Curry, *Sun Rising on the West*, 207.

16. Henry C. "Hank" Smith Papers, Folder 185.

17. *History of Crosby County*, 16; Jenkins, *Estacado*, 250–51.

18. *History of Crosby County*, 16.

19. Henry C. "Hank" Smith Papers, Folder 187; *History of Crosby County*, 18.

20. Henry C. "Hank" Smith Papers, Folder 187.

21. Henry C. "Hank" Smith Papers, Folder 190.

22. *History of Crosby County*, 18, 27.

23. Henry C. "Hank" Smith Papers, Folder 205; Curry, *Sun Rising on the West*, 221.

24. Curry, *Sun Rising on the West*, 221.

25. Wrobel, *Promised Lands*, 11.

26. Henry C. "Hank" Smith Papers, Folder 211; Curry, *Sun Rising on the West*, 220–23.

27. Richard White, *"It's Your Misfortune and None of My Own": A New History of the American West* (Norman: University of Oklahoma Press, 1991), 613.

28. White, *"It's Your Misfortune and None of My Own."*

29. David G. Pugh, *Sons of Liberty: The Masculine Mind in Nineteenth-Century America* (Westport, CT: Greenwood Press, 1983), xv.

30. Matthew Basso, Laura McCall, and Dee Garceau, eds., *Across the Great Divide: Cultures of Manhood in the American West* (New York: Routledge, 2001), 1; Pugh, *Sons of Liberty*, xvi.

31. David Anthony Tyeeme Clark and Joane Nagel, "White Men, Red Masks: Appropriations of 'Indian' Manhood in Imagined Wests," in *Across the Great Divide*, 111–17.

32. Henry C. "Hank" Smith Papers, Folder 46.

33. Wrobel, *Promised Lands*, iii–iv.

34. Henry C. "Hank" Smith Papers, Folder 213; author interview with Georgia Mae Ericson, Crosbyton, TX, November 30, 2000.
35. Curry, *Sun Rising on the West*, 272.
36. Curry, 274.
37. Curry, 273–74.
38. Curry, 276–77.
39. Curry, *Sun Rising on the West*, 276–77; Henry C. "Hank" Smith Papers, Folder 221.
40. Curry, *Sun Rising on the West*, 278.
41. Curry, 281.
42. Curry, 280.
43. Curry, 282.
44. Curry, *Sun Rising on the West*, 216; Henry C. "Hank" Smith Papers, Folder 100.
45. Curry, *Sun Rising on the West*, 281.
46. Henry C. "Hank" Smith Papers, Folder 215; Sosebee, "Henry C. 'Hank' Smith: A Portrait of a South Plains Capitalist," 75–76.
47. Sosebee, "Henry C. 'Hank' Smith: A Portrait of a South Plains Capitalist," 76.
48. Sosebee, 76.
49. Sosebee, 76.
50. Sosebee, 76.
51. Sosebee, 76.
52. Sosebee, 80; Henry C. "Hank" Smith Papers, Folder 217.
53. Randolph B. Campbell, *Gone to Texas: A History of the Lone Star State* (New York: Oxford University Press, 2003), 319–21, 344–45.
54. Henry C. "Hank" Smith Papers, Folder 217.
55. Sosebee, "Henry C. 'Hank' Smith: A Portrait of a South Plains Capitalist," 80–81.
56. Sosebee, "Henry C. 'Hank' Smith: A Portrait of a South Plains Capitalist," 81; Henry C. "Hank" Smith Papers, Folder 218.
57. Sosebee, "Henry C. 'Hank' Smith: A Portrait of a South Plains Capitalist," 81.
58. Sosebee, 83.
59. Sosebee, 84.
60. Sosebee, 81.
61. Sosebee, 82–83.
62. Sosebee, 82–83; Curry, *Sun Rising on the West*, 229–30.
63. Sosebee, "Henry C. 'Hank' Smith: A Portrait of a South Plains Capitalist," 83.

Epilogue

1. Crosby County Pioneer Memorial Museum, Crosby County, Texas, Crosbyton, TX.

Bibliography

Manuscript and Archival Collections

Primary Sources

Bruce Gerdes Collection, Interview Files. Historic Research Center, PanhandlePlains Historical Museum, West Texas A&M University, Canyon, TX.

Bugbee Files. Historic Research Center, Panhandle-Plains Historical Museum, West Texas A&M University, Canyon, TX.

Cox, Paris. Papers. Southwest Collection/Special Collections Library, Texas Tech University, Lubbock, TX.

Estacado, Texas, Collection. Southwest Collection/Special Collections Library, Texas Tech University, Lubbock, TX.

Frank Collinson Collection. Historic Research Center, Panhandle-Plains Historical Museum, West Texas A&M University, Canyon, TX.

J. Evetts Haley Collection, Interview Files. Historic Research Center, PanhandlePlains Historical Museum, West Texas A&M University, Canyon, TX.

Kentucky Cattle Raising Company Records, 1886–1894. Southwest Collection/Special Collections Library, Texas Tech University, Lubbock, TX.

Matador Land and Cattle Company, Dundee Division Records. Southwest Collection/Special Collections Library, Texas Tech University, Lubbock, TX.

Powell, I. P. Papers. Southwest Collection/Special Collections Library, Texas Tech University, Lubbock, TX.

Robert Nail Jr. Foundation Collection. Old Jail Art Center, Albany, TX.

Smith, Henry C. "Hank." Papers. Historical Research Center, Panhandle-Plains Historical Museum, West Texas A&M University, Canyon TX.
Spikes, Joseph. Papers. Southwest Collection/Special Collections Library, Texas Tech University, Lubbock, TX.

Oral Histories

Carpenter, Louis Bassett. Oral Interview. Southwest Collection/Special Collections Library, Texas Tech University, Lubbock, TX.
Carter, Clayton. Oral History Interview. Southwest Collection/Special Collections Library, Texas Tech University, Lubbock, TX.
Crosby County, Texas. Minutes of Commissioners Court, Volume A, Crosbyton, TX.
Ericson, Georgia Mae. Interview. Crosbyton, TX, November 30, 2000. *Government Documents*. Crosby County Deed Records. Crosby County Clerk's Office, Crosbyton, TX.
Shackelford County, Texas. Deed Records. Shackelford County Clerk's Office, Albany, TX.
Shackelford County, Texas, District Court Records, Vol. A. Shackelford County Courthouse, Albany, TX.

Secondary Sources

Books

Arrington, Leonard J., and Davis Bitton. *The Mormon Experience: A History of the Latter-Day Saints.* New York: Knopf, 1979.
Backscheider, Paula R. *Reflections on Biography.* New York: Oxford University Press, 1999.
Basso, Matthew, Laura McCall, and Dee Garceau, eds. *Across the Great Divide: Cultures of Manhood in the American West.* New York: Routledge, 2001.
Baylor, George Wythe. *John Robert Baylor: Confederate Governor of Arizona.* Tucson: Arizona Pioneers' Historical Society, 1966.
Blackbourn, David. *The Long Nineteenth Century: A History of Germany, 1780–1918.* New York: Oxford University Press, 1998.
Boyle, Susan Calafate. *Los Capitalistas: Hispano Merchants and the*

Santa Fe Trade. Albuquerque: University of New Mexico Press, 1997.

Britton, Sarah Ann. *The Early History of Baylor County*. Dallas: Story Book Press, 1955.

Campbell, Randolph B. *Gone to Texas: A History of the Lone Star State*. New York: Oxford University Press, 2003.

Carlson, Paul H. *Texas Woollybacks: The Range Sheep and Goat Industry*. College Station: Texas A&M University Press, 1982.

Cashion, Ty. *A Texas Frontier: The Clear Fork and Fort Griffin. 1849–1887*. Norman: University of Oklahoma Press, 1996.

Conkling, Roscoe P., and Margaret Conkling. *The Butterfield Overland Mail, 1857–1869*. Glendale, CA: A. H. Clark, 1947.

Connor, Seymour V., and Jimmy Skaggs. *Broadcloth and Britches: The Santa Fe Trade*. College Station: Texas A&M University Press, 1977.

Cremony, John C. *Life among the Apaches*. Lincoln: University of Nebraska Press, 1983.

Curry, W. Hubert. *Sun Rising on the West: The Saga of Henry Clay and Elizabeth Smith*. Crosbyton, TX: Crosby County Pioneer Memorial, 1959.

Dale, Edward Everett. *The Range Cattle Industry: Ranching on the Great Plains From 1865 to 1925*. 3rd ed. Norman: University of Oklahoma Press, 1960.

Dary, David. *Entrepreneurs of the Old West*. New York: Alfred A. Knopf, 1986.

———. *The Santa Fe Trail: Its History, Legends, and Lore*. New York: Penguin Books, 2002.

Flores, Dan. *Caprock Canyonlands: Journeys into the Heart of the South Plains*. Austin: University of Texas Press, 1990.

Frazier, Donald S. *Blood and Treasure: Confederate Empire in the Southwest*. College Station: Texas A&M University Press, 1995.

Heathcote, Earl W. "Business in El Paso." In *El Paso: A Centennial History*. El Paso: El Paso Historical Society, 1972.

Hine, Robert V., and Edwin R. Bingham, eds. *The American Frontier: Readings and Documents*. Boston: Little, Brown, 1971.

———. *A History of Crosby County, 1876–1977*. Crosbyton, TX: Crosby County Historical Commission, 1978.

Holden, William Curry. *The Espuela Land and Cattle Company: A Study of a Foreign Owned Ranch in Texas*. Austin: Texas State Historical Association, 1970.

Jenkins, John Cooper. *Estacado: Cradle of Culture and Civilization on the Staked Plains of Texas*. Crosbyton, TX: Crosby County Pioneer Memorial Museum, 1986.

Jones, Fayette Alexander. *New Mexico Mines and Minerals*. Santa Fe: New Mexico Printing, 1904.

Jordan, Terry G. *North American Cattle-Ranching Frontiers: Origins, Diffusions, and Differentiations*. Albuquerque: University of New Mexico Press, 1993.

Kerby, Robert Lee. *The Confederate Invasion of New Mexico and Arizona, 1861–1862*. Tucson: Westernlore Press, 1981.

Knepper, George W. *Ohio and Its People*. Kent, OH: Kent State University Press, 1989.

Larson, Carole. *Forgotten Frontier: The Story of Southwestern New Mexico*. Albuquerque: University of New Mexico Press, 1993.

Limerick, Patricia Nelson. *A Legacy of Conquest: The Unbroken Past of the American West*. New York: W. W. Norton, 1987.

Lomask, Milton. *The Biographer's Craft*. New York, Harper and Row, 1986.

McNeill, J. C., III. *The McNeill's SR Ranch: 100 Years in Blanco Canyon*. College Station: Texas A&M University Press, 1988.

Meleski, Patricia F. *Echoes of the Past—New Mexico's Ghost Towns*. Albuquerque: University of New Mexico Press, 1972.

Miller, Larry L. *Ohio Place Names*. Bloomington: Indiana University Press, 1996.

Nadel, Ira Bruce. *Biography: Fiction, Fact, and Form*. New York: St. Martin's Press, 1984.

Paul, Rodman W. *The Far West and the Great Plains in Transition, 1859–1900*. New York: Harper and Row, 1988.

Pearce W. M. *The Matador Land and Cattle Company*. Norman: University of Oklahoma Press, 1964.

Perry, Richard J. *Western Apache Heritage: People of the Mountain Corridor*. Austin: University of Texas Press, 1991.

Pugh, David G. *Sons of Liberty: The Masculine Mind in Nineteenth-Century America. Westport, CT: Greenwood Press, 1983.

Settle, Raymond W., and Mary L. Settle. *War Drums and Wagon Wheels: The Story of Russell, Majors, and Waddell*. Lincoln: University of Nebraska Press, 1966.

Skaggs, Jimmy. *Prime Cut: Livestock Raising and Meatpacking in the United States, 1607–1983*. College Station: Texas A&M University Press, 1986.

Spratt, John Stricklin. *The Road to Spindletop: Economic Change in Texas, 1875–1901*. Dallas: Southern Methodist University Press, 1955.

Sweeney, Edwin R. *Cochise: Chiricahua Apache Chief*. Norman: University of Oklahoma Press, 1991.

———. *Mangas Coloradas: Chief of the Chiricahua Apaches*. Norman: University of Oklahoma Press, 1998.

Thompson, Jerry, ed. *Civil War in the Southwest: Recollections of the Sibley Brigade*. College Station: Texas A&M University Press, 2001.

———. *Colonel John Robert Baylor: Texas Indian Fighter and Confederate Soldier*. Hillsboro, TX: Hill Junior College Press, 1971.

Timmons, W. H. *El Paso: A Borderlands History*. El Paso: University of Texas at El Paso, 1990.

Tolzmann, Don Heinrich. *The German American Experience*. New York: Humanity Books, 2000.

Trimble, Marshall. *Arizona: A Panoramic History of a Frontier State*. New York: Doubleday, 1972.

Wagoner, Jay J. *Early Arizona*. Tucson: University of Arizona Press, 1975.

Walker, Henry Pickering. *The Wagonmasters: High Plains Freighting Frontier from the Early Days of the Santa Fe Trail to 1880*. Norman: University of Oklahoma Press, 1966.

Weisenburger, Francis P. *The History of the State of Ohio*. Vol. 3, *The Passing of The Frontier, 1825–1850*. Columbus: Ohio State Archaeological and Historical Society, 1941.

White, Richard. *"It's Your Misfortune and None of My Own": A New History of the American West*. Norman: University of Oklahoma Press, 1991.

Wrobel, David. *Promised Lands: Promotion. Memory, and the Creation of the American West*. Lawrence: University Press of Kansas, 2002.

Articles

Anderson, Hattie M. "Mining and Indian Fighting in Arizona and New Mexico, 1858–1861: Memoirs of Hank Smith." *Panhandle-Plains Historical Review* 1 (1928).

Hall, Martin Hardwick. "Captain Thomas Mastin's Arizona Guards, CSA." *New Mexico Historical Review* 49 (April 1974).

Hutto, John. "Mrs. Elizabeth (Aunt Hank) Smith." *West Texas Historical Association Yearbook*, vol. 15, Abilene, TX.

Sosebee, Scott. "Henry C. 'Hank' Smith: A Portrait of a South Plains Capitalist." *West Texas Historical Association Yearbook* 77 (2001).

Tevis, James. "Arizona in the '50s." *True West Magazine*, June–August 1968.

Smythe, Col. R. P. "The First Settlers and the Organization of Floyd, Hale, and Lubbock Counties." in *West Texas Historical Association Yearbook*, vol. 6, Abilene, TX.

Unpublished Sources

Christian, Gama Loy. "Sword and Plowshare: The Symbiotic Development of Fort Bliss and El Paso, 1849–1968." PhD diss., Texas Tech University, 1977.

Index

Abilene, TX, 145
Adams, Sallie Mae. *See* Smith, Sallie
Ake, Felix Grundy, 54; Ake party, 54–56
Albany, TX, 104, 120
Alvaranda, Pablo, 93
Amarillo News, 169, 170
American Express Company, 42
American West: cowboy, 129; cult of masculinity, 157; myth of, 12, 156–59, 171; promotion of, 144–45; shaping factors of, 13–15
Amity Clause, 141
Anchito, Antonio, 69
Anti-Saloon League, 167
Anti-Statewide Prohibition Organization of Texas, 167
Apache, 43, 47–48; conflict with, 43–45, 47, 49–60, 73, 79–82, 154, 155. *See also* Bedonkohe, Chihennes, Chiricahua, Mescalero
Apache Pass, 47, 52
Arizona, 38–40, 41, 42, 43, 44, 47, 50, 53, 54, 60, 61, 62; Confederate acquisition of, 62–63; Confederate Territory of, 73, 76, 78, 80
Arizona Guards, 53–54, 56–60, 61, 66–71, 73, 76, 79–83, 158
A Texas Frontier: The Clear Fork Country and Fort Griffin, 1840–1887 (Cashion), 98
Austrian Succession, War of the, 17

Bailey County, TX, 146
Baiuarii, the, 16
Barella, Ignacio, 69
Bar-N-Bar Ranch, 163
Bavaria, 16–21

Bavarian Succession, War of the, 17
Baylor County, TX, 146, 147–48
Baylor, John Robert, 63–65; Arizona Guards, 56–60; Confederate invasion of New Mexico, 65–73; Indian policy, 78–82; martial governor, 73; relieved of authority, 81–82; retreat from Mesilla, 73–76; trial of, 77–78
Baylor, Lieutenant Colonel John. *See* Baylor, John Robert
Bean, Roy (judge), 49, 155, 174
Bean, Samuel, 49
Bedonkohe, 47, 50, 51
Bericht uber eine Reise nach den westlichen Staaten Nordamerikas (Duden), 23
Bexar County, TX, 146
Big Burro Mountains, 79
Biggers, Don, 107, 108
Birchville, NM. *See* Pinos Altos
Bird, George, 38
Birdwell, John, 111
Blacker, M. V., 123
Blanco Canyon, 11, 16, 108–12, 113–15, 116, 117, 168; agriculture, 138–39; blizzards, 164–65; features of, 118–19; ranching in, 131–38; settlement of 117, 118, 121–28, 143–44, 173–74
Blanco River, 108
blizzards, 164–65
Bonaparte, Napoleon, 17
Boony and Armstrong, 34
booster(s), western, 144–45. *See also* promoter(s)
Borrajo, Father Antonio, 92
Boyle, Elizabeth, 101, 102–3. *See also* Smith, Elizabeth

160 Index

Boyle, James, 101–2, 105, 106
Boyle, Susan Calafate, 29–30
Brazoria County, TX, 125
Brewer, Anton, 54, 55, 56, 76–77
Britton, A. M., 121–23
Brown, Aaron V., 41–42
Brown, Frank J., 151
Brown, Joe, 150
Buel, G. P., 103, 104
buffalo, 87, 118, 119, 132; hunting of, 64, 87–88, 97–99, 105, 108, 113, 115–16, 118, 120, 132; ranch, 107–8
Buffalo Springs, 117
Butterfield, John, 42–43, 89
Butterfield Overland Mail Company, 89

Cajon Pass, 38
California, 37, 38, 39, 41–42, 57, 63, 74, 89, 91, 174
Campbell, Governor Thomas, 167
Campbell, H. H., 121–23, 133, 149
Camp Cooper, 114
Canby, Colonel Edward R. S., 67–68, 74–75, 79
Caprock Escarpment (Caprock), 113, 114, 117–18, 126, 160, 162, 163
Cardis, Luis, 92–94
Carleton, Major James, 84
cattle, 11, 16, 45–48, 54–55, 97, 98, 108, 111, 112, 119–25, 130, 131; foreign investors, 122–23; Hereford, 134, 160, 161, 164, 165; longhorn, 121, 132; open-range, 128–30; Shorthorn, 134; "Texas system," 129
cattlemen, 104, 121, 129, 146; land acquisition, 131–32; sheep raising, 134–38, 160, 161
Central Plains Academy, 127
Chamborino, NM, 70
Chihennes, 49, 51
Chihuahua, Mexico, 29, 33, 63, 80
Chihuahua Trail, 33, 61
Chiricahua, 61; conflict with, 43–44, 47–48, 51–59, 68, 73, 78–81; herders, 47; language, 155, 169
Cimarron cutoff, 32

Civil War, 57, 60, 61, 68, 86, 87, 88, 89, 90, 100, 129, 135, 144, 166
Clark, David Anthony Tyeeme, 158
Clark, John Wharton, 141
Cochise, 44, 52–55, 57–59, 73, 78, 80, 155
Cochran County, TX, 146
Comanche, 63, 64, 114–15, 119, 121. *See also* Kotsoteka, Quahadi
commerce, chamber of, 119. *See also* booster(s), western; promoter(s)
Confederate(s), 61, 62, 63, 72, 75, 76–85; Army, 56–58, 90; invasion of New Mexico, 60, 65–73; scrip, 74
Confederate States of America, 57, 61
Congress of Vienna, 17
Conrad and Rath's, 101, 103, 105, 149
Conrad, Frank, 99, 106, 128
Cooke's Peak, 53
Coronado Expedition, 113
Corralitos, Mexico, 80
Council Bluffs, IA, 34, 35
Council Grove, KS, 31
cowboy(s), 11, 12, 14, 97, 98, 99, 103, 105, 128, 129, 152, 156–57
Cox, Paris, 22, 126–27, 148, 151, 152
Cremony, John, 51, 52
Crosby County Land District, 146
Crosby County News, 151
Crosby County, TX, 14, 16, 113, 121, 122, 123, 124, 146, 147, 148, 149, 150, 151, 153, 159, 161, 163, 164, 166, 173, 174
Crosby, Stephen F., 146
Crosbyton Review, 16, 170
Crosbyton South Plains Railway, 153
Crosbyton, TX, 113, 153, 161, 170, 171, 173
Cross B Ranch (+B), 13, 133, 134, 139, 143, 154, 160, 162, 163, 164, 165, 175
Curry, Hubert, 120, 171

Dallas, TX, 103, 109, 110, 111
Davis, Jefferson, 57, 66
DeLong, Lieutenant, Sidney, 84
DeRider, Frank, 68

Index

Dickens County, TX, 117, 121, 123, 146
Diffenderfer, David, 92
Diffenderfer, Frank R., 92, 93, 94, 96
Dockum, W. C., 117, 121
Dowell, D. M., 103–4
Duck Creek, 114, 117
Dziltanatal (Cooke's Peak), 53, 54

Eastland County School Lands, 108, 121, 124, 132
El Paso Del Norte, 33, 61, 88–89, 92
El Paso, TX, 33, 40, 41, 47, 48, 61, 65, 67, 85, 86, 88–89, 135; Anglo influence, 90; Butterfield Line, 89; Civil War years, 89–90; Hank Smith, 91–96; mail route to Santa Fe, 118
Emma, TX, 151–53, 161, 162–63, 167, 171, 173
Entrepreneurs of the Old West (Dary), 46
Ericson, Georgia Mae, 171
Erie Canal, 26, 27
Espuela Land and Cattle Company, 123
Estacado, TX, 22, 121, 126, 127, 128, 137, 138, 145–48, 150–53, 173
Estelle, Billy, 49

farmer(s), 18, 26, 89, 127, 132, 135, 137, 143, 150, 153; tenant, 18, 139, 162
Fayette County, TX, 63
Floyd County, TX, 123, 126, 146
Fort Bliss, 47, 58, 65–66, 67, 74, 75, 76, 89, 90, 91, 92, 93
Fort Craig, 74
Fort Cummings, 84
Fort Davis, 74, 95
Fort Fillmore, 57, 67, 69, 70, 71, 72, 76, 83
Fort Griffin, 84, 96–112, 120, 127, 167; buffalo trade, 98 – 99, 106, 118; frontier town, 97 – 98; Hank Smith, 99–103, 105–9, 112, 113, 115–16, 117, 118, 128, 130, 149, 154, 162, 164, 173
Fort Laramie, 36
Fort Quitman, 74, 91, 93, 94, 95, 101, 140, 173

Fort Richardson, 115
Fort Stanton, 58, 71
Fort Sumner, NM, 127
Fort Union, 74, 75
Fort Worth Democrat, 99
Fort Worth, TX, 103, 108, 109, 110, 111, 121, 127
Fort Yuma, 38, 43, 74
Franklin, Felix, 148
Franklin, TX, 89. *See also* El Paso, TX
freighter(s), 12, 27, 29, 31, 33, 36, 42, 46, 84, 85, 91, 98, 106; hazards of, 35, 78, 87
French revolutionary wars, 17
Frontier Echo, 106, 118, 119

Gadsden Purchase, 41, 66
Garza County, TX, 123
German Triangle, 25
Gila City, AZ, 38, 39
Gila Ranch, 43, 45
Gila Ranch station, 43
Gila River, 38
Glorieta Pass, Battle of, 82
gold, mining of, 14, 38–39, 41, 48, 49, 51, 60, 61, 67, 76, 86, 112, 140, 155, 174
Gray, T. A., 148
Great Bend of the Arkansas River, 31–32

"Hacienda de Glorieta," 108
Hale County, TX, 146
Hanna, Emily, 64
Hart, Simeon, 61
Hawse, Charley, 84, 115–16, 117, 120, 126, 132–33, 136, 138, 146, 162, 163–64, 168, 171
Helm, Thomas, 53, 68, 69, 79, 81
Hereford. *See* cattle, Hereford
Hicks, W. B., 104
Hicks, W. H., 100
H-L Ranch, 123
Hockley County, TX, 146
Horsbrough, Fred, 126, 133, 160, 164, 165, 169, 170

Horsbrough, Mary, 126
Houston and Great Northern railroad, 125
Hume, H. E., 151
Hunt, Emma, 145, 151
hunter(s), 97, 98, 99, 103, 105, 108, 109, 113, 116, 117–18, 121, 127, 147

immigration, 22–26
Indian Depredation Act, Amity Clause, 141
Indians, extermination of, 78–79, 81, 97
Iowa City, IA, 37
It's Your Misfortune and None of My Own: A New History of the American West (White), 13

Jacobs, John Henry, 116–17, 120
Jackson, Territorial Secretary Alexander, 62

Kansas City, MO, 29, 34, 86, 87
Kansas-Nebraska Act, 33–34
Kelly, Robert, 77
Kent County, TX, 123
Kentucky Cattle Raising Company, 124–25, 131, 137
Key, Postmaster General David M., 128
Kirk, Billy, 163
Kirk, J. W., 163
Kirk, Leila, 163
Kirk, Mary, 163
Knights of the Golden Circle, 65
Knox County, TX, 147–48
Kotsoteka, 114

Lake Erie, 27
Lamb County, TX, 146
Las Vegas, NM, 32
Legacy of Conquest: The Unbroken Past of the American West, The (Limerick), 13
Lemons, John, 38
Leonard, Emma, 125–26
Leonard, Van, 125–26
Livingston and Fargo. *See* American Express Company

Llano Estacado, 113, 115, 116, 118, 127
Lomax, Spottswood, 121, 123
Lomax, Vernon, 163
longhorn, Texas. *See* cattle, longhorn
Louis I, (king), 17
Lubbock County, TX, 146
Lynde, Major Isaac, 69–72

Mackay, Alexander, 122
Mackenzie, Colonel Ranald S., 114–15
Mackenzie, Murdo, 122
Mackenzie Trail, 112, 117
MacWillie, Marcus, 78
Mainz River, 16
Majors, Alexander, 35–36
Mangas Coloradas, 47–48, 49–52, 53–59, 73, 78, 80, 155, 159
Marietta, TX. *See* Estacado, TX
Massey, Barbara, 161–62
Mastin, Captain Thomas J. (Tom), 53, 56, 59–60, 69
Matador Land and Cattle Company, 121–23, 133, 149, 150, 162, 174
Matador Ranch. *See* Matador Land and Cattle Company
Maximilian, 17
McDermett, Agnes, 163
McDermett, Ann, 163
McDermett, Charles, 163
McDermett, Charles Jr., 163
McDermett, Jeanette, 163
McDermett, J. Wilson, 163
McNeill, Captain James, Sr., 125
McNeill, James, Jr., 125
Mescalero, 73, 140–41
Mesilla, NM, 50, 54–56, 58, 60, 66, 68–71, 73–78, 81–85, 86, 91
Mesilla Times, 76, 77
Mesilla Valley, 33, 66, 89
Mexican-American War, 29, 88, 135
Mills, Anson, 89
Mimbres River, 50, 54, 55, 56, 84
Mimbres stage station, 56
Mogollon Mountains, 78, 80
Montgomery Convention, 61
Montgomery, Tom, 123–24
Mormon(s), 36–37

Index

Mormon Campaign, 41
Motley County, TX, 121, 146
Mount Blanco, 128, 173
Mount Blanco, TX, 13, 22, 126, 128, 145–46, 153, 161, 162, 163, 167, 173

Nagel, Joane, 158
Napoleonic Wars, 17
Native Americans, 14, 44, 50, 64–65, 66, 79, 86, 97, 154, 158. *See also* Apache, Comanche
Nebraska Territory, 33–34
Neighbors, Robert S., 64
New Mexico: 29–30, 32, 42, 48, 50, 52–53, 56–58, 89, 91, 135; Confederate invasion of, 60, 61–63, 65–68, 70–83, 89–90
Neyland, Father Malcolm, 171

Occidental Hotel, 105, 112, 115–16
Ohio, 13, 23, 25, 26, 27, 28, 37, 38, 40, 63, 66, 106, 141, 168
open-range, 129, 130, 132, 133. *See also* cattle, open-range
Oregon Trail, 29, 36
Otero, Antonio José, 30
Otero, Manuel José, 30
Otero, Miguel Antonio, 30, 62
Otero, Vincente, 30
Oury, Grant, 78
Overland Mail Company, 41, 42, 45, 89
Overland Mail Route, 41, 44–45, 82
oxen, 30, 31, 35–36, 46, 87

Palmer, Captain I. N., 34
Palo Duro Canyon, 115
Pease River, 121
Pennington, Jack, 43, 45
Pennsylvania, 24–25
Peru, OH, 23–28
Peters Colony, 103
Pinos Altos Mountains, 48
Pinos Altos, NM, 48–54, 56–60, 66, 68 –69, 73, 76, 77, 78–79, 81, 82, 86, 155, 173
Posey, James, 139–40, 166
prohibition, 166–68

promoter(s), 42, 118–119, 143–45, 159. *See also* booster(s), western

Quahadi, 114–15
Quaker(s), 22, 121, 126–27, 128, 138, 145–46, 147, 150–51, 152
Quanah (Parker), 114

Ralls, TX, 153, 171
Reid, Jim, 117
Reily, James, 63
Rencher, Governor Abraham, 62
Robertson, R. L., 81
Rock House, 109, 110, 116–17, 120, 126, 127, 128, 132, 138, 146, 154, 160, 162, 163, 168, 171, 173, 174
Roosevelt, Elliot, 119
Roosevelt, Theodore, 156–58
Rossbrunn, Bavaria, Germany, 16, 18–19, 20, 24, 28
Russell, Majors, and Waddell, 36
Russell, William, 36

Salt Lake City, UT, 36, 37
San Antonio, TX, 63, 65, 75, 83, 88–89, 95, 96, 100, 135, 101
San Bernardino, CA, 37, 38
Santa Fe, NM, 29, 30, 32, 33, 57, 60, 61, 62, 86–87, 89, 118, 127
Santa Fe Trail, 29, 31, 32, 33, 36, 85, 86, 87, 91, 112, 174
Schimerhorn, Judge John, 108
Schmitt, Anna, 18
Schmitt, Anna, 21, 23–27
Schmitt, George. *See* Schmitt, Johann George (George)
Schmitt, George, 26
Schmitt, Heinrich, 16, 19–20, 21, 24, 25, 26, 27–28, 174. *See also* Smith, Hank
Schmitt, Jacob, 21
Schmitt, Johann, 18
Schmitt, Johann George (George), 18–21
Schmitt, Magdelene, 23
Schmitt, Margaret, 21, 23, 169
Schmitt, Margaretta Herman, 19, 20, 21, 23–24

Schmitt, Michael, 18
Schmitt, Peter, 26
Scurry, William Reed, 61
secession, 58, 60, 62–63, 65, 66, 78
secessionist(s), 61
Shackelford County, TX, 96–97, 103, 104, 111, 120
sheep, 54, 133, 134–38, 160, 161
Sherman, General William T., 114
Shorthorn. *See* cattle, Shorthorn
Sibley, General Henry Hopkins, 63, 65, 72–75, 79, 81–83, 90
Silver Falls, 118, 124
Smith, Annie, 162
Smith, Charlie, 110, 111, 116, 120
Smith, Elizabeth, 11, 103, 104, 105–6, 112, 115, 117, 118, 120–21, 145–46, 152, 161–63, 168–71, 173; postmistress, 127–28, 171
Smith, Floyd, 162
Smith, Frank, 162
Smith, George, 106, 117, 133–34, 136–37, 139, 145, 154, 160–62, 164–66, 168, 170
Smith, Hank: 11–15, 16, 18, 21, 22, 28–30, 34, 35, 40, 41–43, 45–48, 86, 91–96, 108–12, 113, 115–20, 143–46, 147, 152–54, 160, 162, 164–65, 167–72, 173–75; Arizona Guards, 58–60, 61; cattleman, 110–12, 130, 132–35, 164–65; Civil War, 63, 66, 70, 73, 76–77, 78–80, 82–85; death of, 170; entrepreneur, 138–42; farmer, 138–39; Fort Griffin, 96–102, 103–7; freighter, 84; historian, 155–59; hog raiser, 165–66; legend, 154; livery stable, 106; Mangas Coloradas, 49, 51–52, 54–55; marriage of, 103: miner, 38–39, 60; nickname, 28; ranch work, 45–46; sheepman, 134–38, 160; tax assessor, 122, 148–50; trader, 88; wagon driver, 36–38, 43; wagon master, 30–33, 86–88; weather recorder, 143
Smith, Henry C. "Hank." *See* Smith, Hank
Smith, Henry Clay "Hank." *See* Smith, Hank

Smith, Leila, 115, 117, 145, 162–63, 164, 171
Smith, Mary, 162, 163
Smith Reuben, 162
Smith, Robert, 136–37, 139, 154, 160, 161–62, 164
Smith, Sallie, 160–61
Smith, Viola, 162
Snively, Jacob, 48
Southern Emigrant Road, 41
Spanish Succession, War of the, 17
Spur Ranch, 123, 160
Spur, TX, 121
SR Ranch, 125, 133, 137, 174
Steele, Colonel William, 82–83
Stonewall County, TX, 117, 136
Stringfellow, R. L., 151
Sutton, Abe, 45–46, 133
Sweeney, Edwin R., 47, 50, 51–52, 80
Swink, George, 146–47, 148

Tasker, Charles, 107–12, 116–17, 119, 120, 121, 128, 132
Taylor, Private Milam, 72
teamsters, 31, 32–33, 100, 109
Texas Brewers Association, 167
Texas General Land Office, 146
Texas: land acquisition, 131–32; public school land grants, 131; railroad land grants, 131
Texas Local Option Association, 167
Texas Mounted Rifles, 65, 67
Tilford, Claude, 124
TM Ranch, 123, 133, 137, 160
Tocqueville, Alexis de, 157
Tucson, AZ, 33, 43, 54, 56, 66, 69, 82, 86–87,
True West Magazine (Tevis), 49
Two-Buckle Ranch, 124–25, 131, 133, 137, 146, 149

Union, 57–58, 66, 68, 69, 70–76, 79, 83–85

Van Dorn, Earl, 65
Vega, Juan, 69

Waddell, W. B., 36
Wadsworth, Attorney General William, 46
Wahl, G. W., 95
Waller, Major Edwin, 68, 77
Warren Wagon Train Raid, 114
Watkins, Roe, 94–96, 100
Watrous, NM, 32
Wells and Company. *See* American Express Company
westerner(s), 13–15, 142, 143, 156–57
Westport, MO, 29, 30, 31, 33
Wheeler, Ernest, 162
Wheeler, J. A., 162
Wheeler, John Jr., 162
Wheeler, Josephine, 162
White Man, 64–65
White, Richard, 13, 156
White River, 113, 143
Wisconsin, 25
Women's Christian Temperance Movement, 167

Yellowhouse Canyon, 117
Yost, Samuel, 62
Young County, TX, 146

Zuloaga, José María, 80